AN INTRODUCTORY GUIDE TO
INDUSTRIAL TRIBOLOGY

Other titles in this series:

An Introductory Guide to Flow Measurement – R. C. Baker
An Introductory Guide to Pumps and Pumping Systems – R. K. Turton

An Introductory Guide to Industrial Tribology

J. DENIS SUMMERS-SMITH

Series Editor
Roger C. Baker

Mechanical Engineering Publications Limited, London

First published 1994

ISBN 0 85298 896 6

© J. D. Summers-Smith

A CIP catalogue record for this book is available from the British Library.

Typeset by Paston Press Ltd, Loddon, Norfolk
Printed and Bound by The Lavenham Press Limited, Lavenham, Suffolk, England

D
621·89
ANI

SERIES EDITOR'S FOREWORD

As an engineer I have often felt the need for introductory guides to aspects of engineering outside my own area of knowledge. MEP welcomed the concept of an introductory series to follow on from my own book on flow measurement. We hope that the series will provide engineers with an easily-accessible set of books on common and not-so-common areas of engineering. Each author will bring a different style to his subject, but some valued features of the original volume, such as conciseness and the emphasis of certain sections by shading, have been retained. The initial volumes are biased towards fluids, but we hope to broaden the scope in the later volumes.

The series is designed to be suitable for practising engineers and technicians in industry, for design engineers and those responsible for specifying plant, for engineering consultants who may need to set their specialist knowledge within a wider engineering context, and for teachers, researchers, and students. Each book will give a clear introductory explanation of the technology to allow the reader to assess commercial literature, to follow up more advanced technical books, and to have more confidence in dealing with those who claim an expertise in the subject.

Denis Summers-Smith has been involved with tribology for over forty years, thirty of these in Imperial Chemical Industries where he was tribology adviser, and ten years in consultancy. His main interest lies in the practical application of tribology with emphasis on increasing the reliability of machines. In recognition of his contribution in this field he was awarded the Tribology Silver Medal of the Institution of Mechanical Engineers in 1975.

This volume provides, in a very readable way, an authoritative introduction to the subject with a wealth of practical insight from someone with a very extensive knowledge of the subject, and covers topics such as friction and boundary lubrication, fluid film lubrication, lubricants, selection of bearing type, and other related topics. The result is not only an introductory guide to tribology, but an immensely useful handbook – even for those with knowledge of the subject.

I would particularly like to acknowledge the encouragement of Mick Spencer, the Managing Editor. We both hope that the series will find a welcome with engineers and we shall value reactions and any suggestions for further volumes in the series.

Roger C. Baker
St Albans

CONTENTS

PREFACE

Tribology is concerned with contacts between surfaces in relative motion. These contacts can be met in the following situations:

- load transmission, as, for example, in bearings, brakes, machining, and forming operations; maintenance of close contact, as, for example, in dynamic sealing;
- guidance and movement, as, for example, with textile spindles, chutes handling solids, digging operations.

Tribology covers all aspects of friction, lubrication, and wear. It is primarily involved with surface interactions, though these may be modified to some extent by the mechanical properties of the underlying material; these surface interactions are controlled by the physical and chemical nature of the surfaces, together with the physical and chemical nature of the surrounding environment.

Tribology thus has an impact on all machines, and some understanding of the subject is necessary if these are to function satisfactorily. The main emphasis in this volume is on the practical aspects of lubrication and wear as they affect the engineer in industry. Because of the wide repercussions it is quite impossible to cover all possible aspects in a short text. However, it is considered that there is no real substitute for some understanding of the physical mechanisms and chemical processes involved. For this reason the first two chapters have been devoted to a discussion of the basic principles underlying the mechanisms of friction and lubrication: this approach is mainly descriptive and the use of mathematics is restricted to the minimum necessary for a proper understanding. Subsequent chapters deal with lubricants, bearing materials, mechanisms and wear, and relate these topics to the basic principles discussed in the first two chapters.

The prime objective in this book is to give the practising engineer a grasp of the basic mechanisms involved in sliding and rolling contacts between solid surfaces, as well as the basic principles of lubrication, without the need for an extensive study of the theory. It is aimed at giving the engineer in industry a 'feel' for the subject, while at the same time providing sufficient practical information to meet his day-to-day needs. No attempt has been made to quote the sources of the information given, but reference is made to those sources where more detailed guidance can be obtained on important practical aspects. Finally a brief bibliography to further reading is appended. This is primarily a list of dirty books from my library shelves; in case there is any misconception, this refers to those books that have grubby thumb marks on the fore edges, showing that they have been constantly referred to, in contrast to the ones that have retained their pristine freshness, having merely sat ornamentally on the shelves. The discussion is limited to the tribology of industrial machinery. IC engines, where in addition to the lubrication of the moving parts, the oil has also to maintain cleanliness in the combusion chamber and valves are not covered. Engine lubrication is a complete subject on its own and is better dealt with separately.

J. Denis Summers-Smith

ACKNOWLEDGEMENTS

I am most grateful to Michael Neale and Geoffrey Taylor, both with a wide interest in tribology, who read the manuscript in draft and offered many useful comments that greatly contributed to the clarity of the text.

NOMENCLATURE

A Area of contact

a Constant

b Width (journal bearing)

C Basic dynamic capacity (rolling bearing)

c_d Diametral clearance (journal bearing)

c_r Radial clearance (journal bearing)

d Diameter

e Eccentricity (displacement of journal centre from bearing centre in journal bearing)

F Frictional force

F_t Tangential force/unit width (gear)

G Duty parameter, $\eta v/p'$, (mechanical seal)

h Oil-film thickness, depth of penetration

K Lloyd's K factor, $\frac{F_t}{d}\left(\frac{r+1}{1}\right)$ (gear)

k Wear coefficient

L Life

l Wear depth

N Frequency of rotation (rev/min)

n Frequency of rotation (rev/s)

n_c Critical frequency of rotor supported on knife edges

n_c' Critical frequency of rotor supported on bearing oil films

O Bearing centre

O' Journal centre

P Equivalent radial load (rolling bearing)

$'PV'$ Parameter expressing severity of operating conditions in mechanical seal (P = pressure drop across seal, V = mean face velocity)

p Pressure, specific bearing load on journal bearing (W/bd)

p' Load/unit width (mechanical seal)

p_m Mean yield pressure

q Shear stress

R_a Surface roughness (rms)

R_q Surface roughness (cla)

r Gear ratio

S Sliding distance

s Shear strength

T Temperature

t Time

V Volume wear

v Velocity

W Load

x Exponent in rolling bearing life equation

ε Eccentricity ratio, e/c_r (journal bearing)

η Dynamic viscosity

λ Specific film thickness, h_{min}/σ

μ Coefficient of friction

ν Kinematic viscosity

ρ Density

σ Composite surface roughness (rms), $\frac{1}{\sigma} = \frac{1}{R_{q1}} + \frac{1}{R_{q2}}$ (R_{q1} and R_{q2} are roughnesses of surfaces 1 and 2)

ψ Attitude angle, angle between load line and line of centres (journal bearing)

to Michael J. Neale, a doughty torchbearer for the application of the principles of tribology to engineering practice.

CHAPTER 1

Friction and Boundary Lubrication

Friction is the resistance to motion between surfaces in solid contact. It depends on:

- the load;
- the chemical condition of the surfaces, in particular the presence of contaminant oxide and adsorbed vapour films.

Friction is reduced by the addition of *boundary lubricants* – chemicals that react to form an additional adsorbed layer on the surface – and by *solid lubricant* films.

This first chapter deals with contacts between solids and, in particular, what happens when such surfaces are moved relatively to each other.

1.1 THE NATURE OF SOLID SURFACES

The surfaces of solids, except for those produced by very special techniques, are not truly flat. The surfaces of crystals have steps on them that may be several atomic diameters in height; the surfaces of solids produced by more normal engineering techniques, for example, casting, machining, plating, or fracturing, have surface roughnesses several orders of magnitude greater than this.

The nature of such surfaces is of considerable importance in understanding the mechanisms of friction and lubrication, and it is thus desirable to start by briefly reviewing one of the methods of describing surfaces that is widely used in this connection; this is based on obtaining a line transect of the surface roughness by drawing a fine stylus across it. Clearly this is limited in resolution by the radius of the tip of the stylus and, if the surface has a 'lay' resulting from the method of manufacture, it is necessary to take transects in more than one direction to obtain a complete picture of the surface.

1

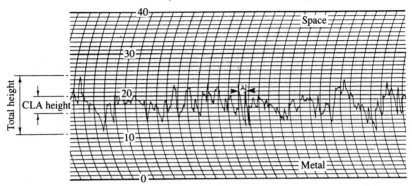

Fig. 1.1 Typical surface finish trace; vertical magnification ×50 (photographically reduced to 70 percent). (Source: *An Introduction to Tribology in Industry,* 1969 (Machinery Publishing Company) p. 14)

Figure 1.1 shows a typical trace obtained from a machined surface with a stylus instrument. The roughness, or departure, of this line from a hypothetical flat line (the centre line) separating the solid and air, can be expressed by an rms or arithmetical average value of the departure of the trace from the centre line. The arithmetical average value, known as the centre line average or cla value and designated by R_a is most commonly used in Europe. (The rms value, designated by R_q, is more widely used in America; it is normally about 10–20 percent greater than the R_a value.) The R_a value obviously does not give the whole story, for example, it says nothing about the spacing of the peaks; however, taken together with a description of how the surface has been prepared it does give a simple description of the surface texture that is sufficient for most practical purposes. A point of interest about the R_a value is that it is about a quarter of the maximum peak-to-valley height for turned surfaces and approximately an eighth for ground and lapped surfaces. Further information on the measurement of R_a and alternative ways of assessing surface texture are given in British Standard BS 1134.

A point to be noted from typical surface traces, such as that shown in Fig. 1.1, is that in order to have a length of surface sufficient to give a representative sample, the vertical movement of the stylus is magnified with respect to the horizontal (usually by 20–40 times). This means that surface roughness traces do not give a genuine physical picture of real surfaces,

Table 1.1 Typical R_a values obtained in different manufacturing processes

	μm	μin
Turned	1–6	32–250
Coarse ground	0.4–3	16–125
Fine ground	0.2–0.4	16–125
600 emery	0.05–0.1	2–4
Lapped	0.02–0.05	1–2

which rather than saw toothed as suggested by the trace, are in fact undulating with slopes typically less than 10 degrees.

Some typical values of surface roughness obtained by different manufacturing processes are given in Table 1.1.

1.2 FRICTION

Now let us consider what happens when such surfaces slide over each other. When one body is lowered on to another, the first contact occurs between two opposing asperities. The materials at the contact deform under the high stress, initially elastically and then plastically when the yield point is exceeded. This allows the two surfaces to approach, more asperities come into contact, and the process of elastic deformation and plastic deformation continues until equilibrium is reached when the area of contact, A, is sufficient to support the load, W. This means that the area of contact is proportional to the load and inversely proportional to the 'yield pressure', p_m, of the softer material. ('Yield pressure' is the pressure that ductile materials subjected to localized plastic deformation can support; it is approximately constant and is equivalent to the indentation hardness.)

$$A = W/p_m$$

REAL AREA OF CONTACT

In practice the real area of contact is substantially independent of the surface finish and normally less than 1 percent of the apparent area.

Using this concept of the contact between surfaces, Bowden and Tabor from Cambridge University, UK, developed a theory to account for the friction that occurs when one surface slides across another. This is based on the idea that welds form at the contacts and the friction arises from the force required to shear these welds. This adhesion theory of friction leads to the following expression for the frictional force, F

$$F = A.s$$

where A is the sum of the areas of the microcontacts and s the mean shear strength of the contacts, i.e., the tangential force/unit area required to shear the contacts; by substituting for A we get the following expression for the frictional force

$$F = W.s/p_m$$

This relatively simple picture gives a satisfactory description of the two laws of friction, which were first enunciated by Leonardo da Vinci and later rediscovered by Amontons with whose name they are now associated.

AMONTONS' LAWS OF FRICTION

(1) The frictional force is independent of the apparent area of contact.

(2) The frictional force is proportional to the load.

Bowden and Tabor in their analysis of friction also pointed out that, if the difference in hardness of the two materials is very high, the harder material will plough a groove in the softer material. Thus the frictional force should be made up of two terms: an adhesion term and a ploughing term. In many practical situations the ploughing term can be ignored, though it should be noted that it again depends on the yield pressure of the softer material.

According to Amontons' second law the ratio between frictional force and applied load is a constant; this constant is known as

the coefficient of friction, μ

$$\mu = F/W = s/p_m$$

As the yield pressure is generally about 5 times the critical shear stress, the coefficient of friction would be expected to be about 0.2. However, this assumes that yielding in compression and yielding in shear can be treated independently. This is not the case and it can be shown that the yield criterion for a junction subject to a normal stress, p_m, and a tangential stress, s, has the following form

$$p_{m^2} + as^2 = \bar{p}_{m^2}$$

where a is a constant with a value of about 10 and $\bar{p}_m$ is the static contact pressure. When the surfaces are loaded normally, there is no tangential stress $s = 0$ and hence $p_m = \bar{p}_m$. However, as soon as a small tangential stress is applied the equality in the equation can only be satisfied if p_m decreases. It has previously been shown that the area of contact is given by $A = W/p_m$; hence if p_m decreases, A must increase, this occurs by the surface sinking together. It would appear that with increasing tangential stress, this process would continue almost indefinitely, being limited ultimately only by the geometry of the contact, and the coefficient of friction would become extremely high. This does not happen because real surfaces are not clean – they are covered by oxide films and layers of adsorbed gases; these interfere with junction growth and allow relative sliding to occur at some value less than the critical shear strength.

IMPORTANCE OF CONTAMINANT FILMS

Surfaces exposed to normal atmospheric conditions are covered by films of oxide and adsorbed vapour. In normal circumstances these films prevent large-scale welding and allow sliding to take place without gross damage.

For example, if the value is 0.9 of the critical shear strength, the contact area increases by about 2½ times before shear takes place, and the coefficient of friction has a value of approximately 0.6 (Fig. 1.2). With perfectly clean surfaces, e.g., the surfaces of noble metals that do not form oxide films or of other metals in vacuum, this restriction is removed and, in fact, gross welding occurs between the surfaces and it becomes meaningless to speak of a coefficient of friction. This is obviously a very real problem in space technology, but it can also occur in mechanisms that have to operate

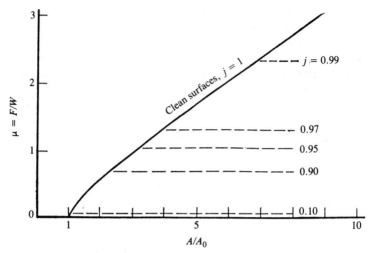

Fig. 1.2   Dependence of the coefficient of friction and increase in the area of contact with the application of tangential force of the fraction (j) of the critical shear stress at which sliding occurs. A_0 = static area of contact; A = area of contact when sliding occurs. (Source: *An Introduction to Tribology in Industry*, 1969 (Machinery Publishing Company), p. 17)

in very pure non-oxidizing gases (e.g., nitrogen and hydrogen), though equally it is used beneficially in friction welding, when the heat generated by the friction desorbs the protective films.

It can be seen that another contradiction to Amontons' second law occurs in cases in which the loads are not high enough to disrupt the oxide film. For example, a very adherent oxide film is formed on copper surfaces; at low loads it remains coherent, giving $\mu = 0.4$, whereas at high loads the film is disrupted and μ rises to about 1.6. With aluminium, however, the match between the crystal structure of the oxide film and that of the parent metal is not as good and, even at very light loads, the film is disrupted and the coefficient of friction is 1.2, irrespective of the applied load. It is this property of the oxide film on aluminium surfaces that makes the cold welding of aluminium possible.

The understanding of the nature of the mechanism of friction and of the effect of these surface contaminants has led to the concept of deliberately applying thin films of low shear strength materials to the surfaces. If the

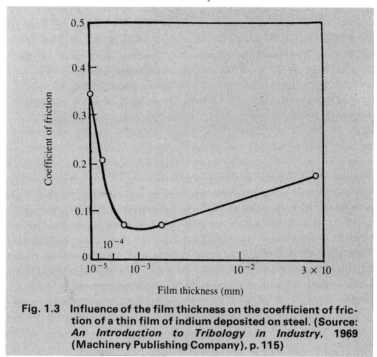

Fig. 1.3   Influence of the film thickness on the coefficient of friction of a thin film of indium deposited on steel. (Source: *An Introduction to Tribology in Industry,* **1969 (Machinery Publishing Company), p. 115)**

film is sufficiently thin, then it will be supported by the substrate material and the area of contact determined by the yield strength of the substrate, while the shear strength of the junctions will be that of the softer coating. For example, films of indium or white metal on steel give a minimum coefficient of friction of 0.06 at a thickness of 10^{-3} to 10^{-4} mm (Fig. 1.3).

This knowledge of the effect of surface phenomena during sliding is of considerable technological significance. Aluminium and titanium are unsatisfactory materials for mating screw threads, where high contact stresses lead to seizure. Noble metals like gold, which do not form oxide films, are even more unsatisfactory. Further, it gives an explanation of the engineer's dislike of running two similar materials together: the more similar the crystal structure of the two surfaces, the more readily will welds form and the greater the likelihood of surface damage and seizure.

CHOICE OF METAL PAIRS FOR SLIDING CONTACTS

Avoid:	gold	–	no oxide film formed
	titanium (aluminium)	–	oxide films are brittle and less effective in preventing welding
	like pairs	–	more prone to welding than dissimilar pairs

So far mainly metallic contacts have been considered in which plastic flow and cold welding are familiar concepts. Experiments show that similar values of the coefficient of friction are obtained between non-metallic surfaces or at a mixed contact between a metal and a non-metal where these ideas appear less acceptable, particularly in the case of brittle solids.

The coefficient of friction between, for example, two crystals of rock salt, is 0.8 – very similar to that of steel on steel. The explanation is that the material at the contact is under very high hydrostatic pressure and it can be shown that rock salt under these conditions behaves plastically and can withstand shear stresses many times higher than it can under more normal conditions. Moreover, under plastic flow conditions, strong adhesion occurs between the crystals giving something very analogous to cold welding between metals.

Contaminant surface films also play a similar role with non-metallic materials. For example, the coefficient of friction between diamond surfaces is 0.5; in a vacuum the contaminant surface is desorbed and the coefficient of friction rises to unity. However, gross seizure does not occur as it does with metals, presumably because the deformation at the contact is purely elastic and junction growth by plastic deformation under combined normal and tangential stresses does not occur.

Rigid polymers present rather a different picture. These materials have visco-elastic properties and thus the area of contact is no longer proportional to load, since it is affected by the geometry of the contact and the length of time the load has been applied. Unlike the case of metals, where the area is directly proportional to the load, with plastics it is proportional to some power less than one; hence the coefficient of friction decreases as the load is increased.

The plastic polytetrafluoroethylene (ptfe) deserves special mention.

PTFE

ptfe has a coefficient of friction of about 0.05, the lowest of all known solids.

(Ice also has low friction, but this arises because under pressure the ice melts to provide a liquid lubricating film. As polar explorers know to their cost, this does not occur at very low temperatures simply because the ice does not melt.) The friction of ptfe is much less affected by adsorbed gases and vapours and remains low over the temperature range $-200°C$ to $300°C$. A completely satisfactory explanation for the low friction of ptfe has not been given, though it probably arises from the anisotropic yield properties of the ptfe crystallites. With the exception of ptfe, coefficients of friction for most engineering materials in a clean and dry state lie in the range 0.1 to 1.0.

1.3 BOUNDARY LUBRICATION

Under the conditions of sliding that have been discussed up to now, the loss of energy resulting from friction is comparatively high and there is the risk of damage to the surfaces. In lubrication, a third component is added to the system in order to reduce the friction and ideally to eliminate damage. The aim in lubrication is to separate the surfaces by a continuous fluid film. Friction losses then only arise from the viscous resistance of the fluid film and, as the surfaces do not touch, there should be no wear. However, even when the film breaks down and solid contacts occur in the presence of a lubricant, there is still a reduction in coefficient of friction and surface damage compared with unlubricated sliding (Fig. 1.4). This is the field of boundary lubrication.

Boundary lubrication can be demonstrated by the deposition of a single monolayer of soap on a steel surface. (A soap is the metal salt of a fatty acid, e.g., stearic acid reacts with the oxide film on steel to form iron stearate, i.e., a soap.) Such a film gives a coefficient of friction of about 0.1 compared with 0.6–0.8 for an unlubricated steel surface. The film rapidly

Fig. 1.4   Wear tracks of stainless steel hemisphere on stainless steel plate, unlubricated (left), lubricated (right), with medium viscosity mineral oil. (Source: *An Introduction to Tribology in Industry*, 1969 (Machinery Publishing Company), p. 20)

BOUNDARY LUBRICATION

Contaminant surface films protect surfaces against damage during sliding. Boundary lubrication extends the process by providing additional surface films that form by the reaction between a fluid lubricant and the oxide film present on the surface.

wears away with repeated traverses, though if a multilayer film is deposited a reasonably stable low friction surface can be produced (Fig. 1.5). The thickness of these films is about 10^{-9} m. Similar effects occur with long-chain hydrocarbon molecules, but the best effects are produced with polar molecules (i.e., molecules that are chemically active at one end) and particularly with those molecules that can react chemically with the surface and can pack closely together to form a coherent film rather like a carpet pile. For example, a long-chain fatty acid in solution in a hydrocarbon forms a very effective lubricating film by reacting with a metal surface to form a soap *in situ*; this has the advantage that if the boundary film is mechanically destroyed, it will reform once the contact is broken and the surface re-

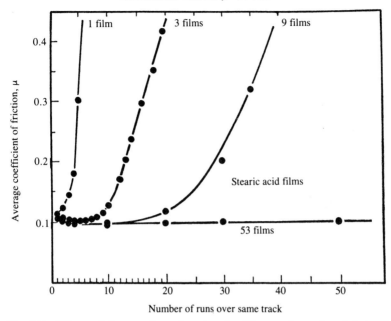

Fig. 1.5   **Effect of the number of molecular films on boundary lubrication.**
(Source: *An Introduction to Tribology in Industry*, 1969 (Machinery
Publishing Company), p. 21)

wetted (Fig. 1.6). Good boundary lubrication is obtained with very dilute
solutions of fatty acids in mineral oils. Mineral oils by themselves have reasonable boundary lubricating properties, not from their hydrocarbon
molecules, but from small amounts of acidic material formed by oxidation.
A certain caution has to be exercised in the formulation of lubricants: if
antioxidants are added to prolong the life of the oil, the acidic degradation
products will not be formed and it may even be necessary to compensate
for this by incorporating a small amount of an 'oiliness' additive.

Long-chain paraffins provide some boundary lubrication, but they are
not as effective as the more reactive materials. Evidence indicates that the
effective lubricant is a solid film: for example, the paraffins cease to have
effective boundary lubricating properties when their melting point is
exceeded. In the same way lauric acid ceases to be effective on plantinum at

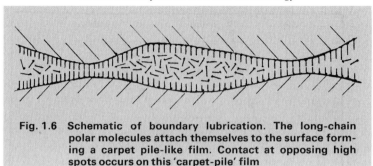

Fig. 1.6 Schematic of boundary lubrication. The long-chain polar molecules attach themselves to the surface forming a carpet pile-like film. Contact at opposing high spots occurs on this 'carpet-pile' film

44°C, its melting point; on copper, on the other hand, with which it reacts to form copper laurate, it continues to lubricate up to 110°C, the melting point of the copper laurate soap.

LIMITS TO BOUNDARY LUBRICATION

Boundary lubrication is only effective on normal engineering materials to about 120°C when the adsorbed films melt or desorb.

For fatty acids to react, it is necessary that the metal should be covered with an oxide film and that moisture should be present. In normal atmospheric conditions this presents no problem, as the small amounts of water and oxygen that dissolve in the lubricating oil are sufficient. In their complete absence, however, lubrication can be a problem. For example, ball bearings operating in a pure helium atmosphere (applications for this occur in nuclear engineering) have a very short life. This can be remedied by incorporating an oxidizing agent in the lubricant.

Another experience showing the importance of the necessary conditions to form surface films concerned the lubrication of some reciprocating compressors handling very pure nitrogen (oxygen and water content less than 2 ppm). The cylinder lubricator for these compressors was located in the crankcase. The first compressors operated satisfactorily, but some new machines were installed, and as an extra precaution to prevent contamination of the nitrogen by leaks from the crankcase, the crankcases were blanketed with the pure nitrogen. After about 500 h on line, when presumably any water and oxygen left in the system at installation had been used up, the piston rings scuffed badly. This happened several times before the cause was identified. The problem was cured by removing the lubricant supply for the cylinders to a polyethylene container in contact with the atmosphere; where the oil, kept saturated with water vapour and oxygen by diffusion through the porous polyethylene, provided satisfactory boundary lubrication for the cylinders.

Problems also occur if the surface is unreactive; for example, stainless steel and titanium are difficult to lubricate under boundary lubrication conditions and have to be avoided in bearings. The coefficient of friction between titanium surfaces is unaffected by normal lubricants and they are particularly prone to galling. Even tightening up a titanium nut on a titanium thread will cause it to seize so that it cannot be taken off without extensive damage. Cast iron, low alloy steels, copper alloys and white metals react readily with most lubricants to form boundary lubricating films, so it is no surprise that they are widely used in tribological mechanisms, e.g., bearings, gears, piston rings. Aluminium and its alloys are much less satisfactory under boundary lubricating conditions and are best avoided for bearing surfaces.

The coefficient of friction under boundary conditions would be expected to arise in a similar way to that of a thin soft metal film on a hard substrate, the shear strength of the boundary film replacing that of the soft metal. A simple calculation for a copper stearate film on copper gives a coefficient of friction of 0.05, whereas it is found experimentally to be 0.1. This increase could be explained by an increase in the shear strength of the soap film under pressure, but it is more probably the result of penetration of the film by asperities on the metal surface. If it is assumed that there is metallic contact over a fraction x of the total area of contact, then friction is given by

$$F = A\{xs_\mathrm{m} + (1 - x)s_\mathrm{s}\}$$

where s_m is the shear strength of the metal and s_s the shear strength of the soap. For the case of copper stearate on copper, this suggests that the area of metallic contact is one or two percent. The most important function of the boundary lubricating film is to prevent growth of the small metallic contacts under the combined normal and tangential stresses set up during sliding. It must be appreciated, however, that such metallic contacts do occur and that there will always be some wear under normal boundary lubrication conditions.

Amontons' laws of solid friction are obeyed under conditions of boundary lubrication. However, with unlubricated sliding, the coefficient of friction is independent of the speed of sliding: under boundary conditions it may either rise or fall with increasing speed. In the former case this has little significance: in the latter, however, it can have important practical consequences. Consider the case of a simple slider on a moving plate restrained by a spring. As the plate is moved away, the tension in the spring increases until it reaches the value of the static friction and relative sliding occurs. In the case of a falling coefficient of friction with speed, this leads to a decrease in friction, and, as the friction decreases, so the relative speed between the surfaces increases; the spring passes through its neutral point and goes into compression until it stops the relative movement of the slider. Sliding now ceases until the spring tension equals the static friction once more, and the process is repeated. This gives rise to an intermittent motion that is termed 'stick–slip' and it is this mechanism that causes the squeaking of door hinges and the squealing of brakes. Stick–slip sliding is highly undesirable in machining operations as it gives rise to chatter marks on the workpiece. Machine-tool lubricants are specially compounded so that the coefficient of friction under boundary lubrication conditions increases with speed so that stable sliding conditions are maintained.

ANTIWEAR AND EXTREME PRESSURE ADDITIVES

Chemically-active materials added to the lubricant react with oxide surface films at temperatures above the limit of boundary lubrication.

Application is in heavily-loaded steel–bronze and steel–steel contacts.

To overcome the problem set by the breakdown of the boundary film at its melting point, special additives can be incorporated in the lubricant. These are the so-called load-carrying and extreme-pressure (ep) additives. Their function is to react with the surface at high temperature, when the boundary film is desorbed, to provide an easily-sheared layer that prevents welding of the two surfaces. Unlike boundary additives, most of these additives do not reduce the coefficient of friction at the contact, but merely function to prevent the welding that would otherwise result in extensive surface damage. These additives are normally organic compounds containing either sulphur, phosphorus, or chlorine. Such materials are potentially corrosive and the formulation of ep oils requires the correct balance of chemical activity: this means that ep additives tend to be specific to the bearing materials. For example, certain additives that are used with steel are so chemically aggressive that they attack copper at normal ambient temperatures and thus cannot be used in systems where there are copper alloys. In effect, ep lubrication is a controlled sacrificial corrosion. It should be noted that load-carrying and ep additives still depend on the presence of oxide films for reaction with metal surfaces, hence they do not provide a solution to the problem of lubrication in non-oxidizing environments.

1.4 SOLID LUBRICATION

An alternative to boundary lubrication is lubrication by solids. The best solid lubricants give coefficients of friction of the same order as in boundary lubrication, though the mechanism is quite different.

SOLID LUBRICANTS

Low friction solids, such as graphite, molybdenum disulphide, and ptfe bonded to a surface provide a low friction film akin to that obtained under boundary lubrication.

Solid lubricants commonly have layer lattice crystal structures with a weaker interatomic bonding between the layers than within them and it is presumed that if the crystals can be bonded to a surface (either mechanically or by interatomic forces) then shear takes place between the layers of the crystal giving low friction.

Graphite is only effective in the presence of water or other contaminants and it has been suggested that the contaminant molecules are dissolved interstitially between the carbon layers giving a further reduction in inter-

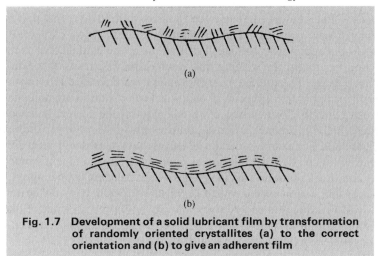

(a)

(b)

Fig. 1.7   Development of a solid lubricant film by transformation of randomly oriented crystallites (a) to the correct orientation and (b) to give an adherent film

layer strength. Some support for this is given by Savage, who showed that the minimum concentration of lubricating vapour in the atmosphere necessary to allow effective lubrication by graphite is inversely proportional to the molecular size. Practical use has been made of this in predicting the suitability of graphite piston rings in unlubricated compressors in terms of the concentration of potential lubricating vapours in the gas being compressed.

Molybdenum disulphide, on the other hand, does not require the presence of adsorbed vapour films and, in fact, its friction increases in the presence of water or organic vapours.

If an effective film of a solid lubricant is to be formed on the surface, the crystals must approach to within atomic distances and be correctly oriented (Fig. 1.7). Clearly these requirements will be impeded if chemisorbed liquid lubricant films are present: for the effective application of solid lubricants not only are clean surfaces required, but also some mechanical action to bring the crystals into the correct orientation for adsorption; for example, buffing techniques can be used.

The use of lubricating solids provides an extension to more normal lubricating techniques. They can be used at temperatures outside the range of oils and greases, both at low and high temperatures. Molybdenum disulphide can be used up to about 300°C, graphite up to 500°C in oxidizing environments (e.g., in air) and up to 1,000°C in non-oxidizing environments. Solid lubricants are also useful as assembly compounds, thread release compounds, where they remain trapped in the contact area, where oils and greases tend to slowly migrate away, and in situations where an oil or grease could give a risk of contamination.

It is worth noting that the adsorbed vapour and oxide films, that are present on all normal engineering surfaces exposed to atmosphere, in fact act as a form of boundary and solid lubrication. Without them many surfaces would seize and grind to a halt.

1.5 FRETTING

If a stress is transmitted across two surfaces held together entirely by mechanical means some slip will tend to occur at the interface. Because of the small movements involved, any lubrication is in the boundary regime and access of fresh lubricant is prevented by the close fitting of the components. In joints subjected to reciprocating stresses, wear occurs because of the repeated small movements; as the wear products created cannot escape they contribute to the process. This can lead to loss of interference fits, while seizure may be caused in the case of sliding fits. Fretting wear results in the formation of pits, which, by reducing the fatigue strength of the component, can result in fatigue failure.

FRETTING

Fretting is defined as the wear of close fitting contacts caused by limited oscillating movement that prevents access of lubricant.

Wear products cannot escape and contribute to the wear process.

Remedies

| Interference fits | – increase interference or interpose film in which movement can be absorbed elastically |
| Sliding fits | – coat surfaces with solid lubricant film |

In engineering practice iron and steel are most commonly affected and in normal atmospheric conditions the iron wear products oxidize, giving reddish-brown deposits. This has given rise to the term 'fretting corrosion', though it should be noted that, while corrosion aggravates the situation, it is not a necessary part of the process. In fact, all materials are liable to fretting, and it is worth pointing out that, with soft materials like plastics operating against a metal surface, frequently it is the metal surface that wears most, the reason being that the oxidized wear particles from the metal surface, which are harder than the parent metal, are embedded in the plastic, which then acts like a lap.

Two cases of fretting must be distinguished. First, there is fretting at interference fits subject to varying stresses (e.g., discs on shafts, roller bearing races in housings). Secondly, in mechanisms where the parts are normally stationary, but may be required to move from time to time (e.g., trip mechanisms, secondary ptfe wedge-ring seals in mechanical seals), the movement to cause the fretting results from vibration. The preventive measures taken in these two cases are quite different.

With interference fits, sliding may be prevented by increasing the interference, though if this is unsuccessful the fretting damage will be more severe. Alternatively, it may be stopped by modifying the contact so that the movement is absorbed elastically, rather than by sliding, by interposing a layer of rubber, or plastically, by coating one of the surfaces with a soft metal. Indium, which recrystallises at normal atmospheric temperatures, is a suitable material for this as it does not work harden and break up. For example, problems with fretting of rolling bearings in their housings have been cured by putting a thin flash of indium on the outer race seating.

The best action with intermittent sliding mechanisms is to coat with a solid lubricant film. This will not completely prevent fretting, but is effective in delaying its onset. Both graphite and molybdenum disulphide can be used (the latter appears to be more effective) and should preferably be applied as a resin-bonded film to a prepared surface. Phosphating gives the best results with ferrous surfaces; anodizing can be used with aluminium surfaces. Mechanical roughening gives some benefit, but is less effective than the chemical treatments. A deposited coating of alumina has been effective in combating fretting wear at the contact between the shaft and ptfe secondary seal in a mechanical seal, presumably because no transformation of the alumina fretting wear products takes place and, being no harder than the original surface, they do not cause wear even if embedded in the plastic.

CHAPTER 2

Fluid Film Lubrication

The friction under conditions of *boundary lubrication* is unacceptably high for many applications.
In the majority of mechanisms the moving contacts are separated by fluid films
either pressurized externally – *hydrostatic lubrication*
or by the relative motion   – *hydrodynamic lubrication*

Hydrodynamic lubrication depends on

- the geometry of the contact;
- the load between the surfaces;
- the relative motion;
- the viscosity of the fluid in the film.

The nature of lubricant films is best described by the *specific film thickness*, that is the ratio of the lubricant film thickness to the surface roughnesses.

In the first chapter we considered what happens when surfaces slide over each other in solid contact. Even in the presence of boundary-lubricating films, friction is still comparatively high and some wear of the surfaces is liable to take place.

The prime aim in lubrication is to separate the faces completely by a fluid film: this eliminates wear and considerably reduces the friction losses.

In order to separate the faces by a fluid film it is necessary to generate pressure in the film. This can be done externally or through the relative motion of the surfaces; the first of these is known as hydrostatic lubrication, the second as hydrodynamic lubrication.

2.1   HYDROSTATIC LUBRICATION

Hydrostatic lubrication requires a pump to generate the pressure; this makes for a complex arrangement and for this reason is less frequently met in industrial machines. Hydrostatic lubrication has the advantage that, once the surfaces have been separated by a film, only very small forces are required to move them relatively. This can be of benefit where it is necessary to position large masses very accurately, for example, a workpiece on the bed of a machine tool or a large astronomical telescope. Similarly, in a machine with a very heavy rotor, hydrostatic lubrication can be used to lift the rotor before starting (hydrostatic jacking), thereby reducing the risk of damage under boundary lubrication conditions before hydrodynamic lubrication takes over on running up the machine to speed. Secondly, it is possible to precisely control the stiffness of the film, a feature that can be of value in a high-speed grinding machine where the grinding wheel must be kept steady.

2.2   HYDRODYNAMIC LUBRICATION

Self-generation of a lubricant film by relative motion, however, normally gives a much simpler arrangement, and is the lubrication mechanism most commonly found in industrial machines.

Pressure in the lubricant film in hydrodynamic bearings is generated by 'wedge action' – the relative movement of the surfaces dragging the lubricant into a decreasing space.

2.2.1   Sliders and thrust bearings

The essential feature of the generation of pressure through shearing a viscous lubricant is that there should be a restriction. This is provided in a simple thrust bearing by having the pad tilted with respect to the running collar, forming a wedge shape so that there is a diminishing thickness of film in the direction of motion. No slip occurs between the surface of a solid and a fluid, so fluid is drawn into the wedge and pressure

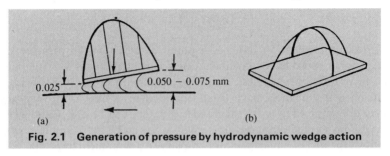

Fig. 2.1   Generation of pressure by hydrodynamic wedge action

is built up. If the fluid is drawn into the wedge faster than it can escape sideways, sufficient pressure is created to float the surfaces apart and a hydrodynamic bearing is produced. The pressure must be ambient at the two ends so the pressure distribution takes the form of a hump, as shown in Fig. 2.1(a). In three dimensions it has a bowler-hat shape, the pressure also falling to ambient at the sides (Fig. 2.1(b)). The integration of the fluid pressure over the pad is equal to the load on the pad, and the system is in equilibrium. The higher the velocity of sliding and the higher the viscosity of the lubricating fluid, the greater the build-up of pressure and the greater the load that can be carried; for the same load, the greater the thickness of the supporting film that is generated.

VISCOSITY

The only fluid property that is of importance in hydro-dynamic lubrication is the viscosity (the resistance to flow).

Most industrial lubricants are liquids, but it should be noted that gases can also be used, though the load-carrying capacity will be less because of their lower viscosity.

There is an optimum pad slope for pressure generation corresponding to the velocity and the viscosity of the lubricant. This can be realized by allowing the pad to tilt to accommodate itself to the operating conditions. In the original designs developed by Michell in Australia and Kingsbury in America, the pivot point was offset slightly towards the trailing edge to develop the maximum lift and ensure stability. This, however, means that the bearing can only be used for one direction of rotation. It has been

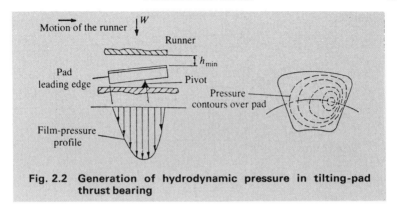

Fig. 2.2 Generation of hydrodynamic pressure in tilting-pad thrust bearing

found, however, that centrally-pivoted pads are equally effective, the wedging action being provided by thermal effects that distort the pads so that they take on a slight crown. Bearings with centrally-pivoted pads have the advantage that they operate equally well in both directions of rotation. Figure 2.2 shows the pressure profile on a single pad, together with the pressure contours over the surface of the pad.

THRUST BEARING TYPES

Rayleigh step	– step height	20–50 μm
Tapered land	– $h_{(inlet)} : h_{(outlet)}$	2:1 to 3:1
Michell tilting-pad	– offset step pivot ca. 60 percent from inlet central step pivot	
Kingsbury tilting-pad	– central button pivot	

It should not be thought, however, that this tilting action is essential for the operation of thrust bearings. It provides the optimum capacity for a range of operating conditions, but there are fixed-pad bearings, known as 'tapered-land' bearings, in which the pad is scraped or machined to provide a wedge approximately appropriate to the particular imposed conditions. In fact, it is very easy, once the principle is recognized, to substantially provide all that is required in load-carrying capacity for moderate duties: rarely are the designs critical.

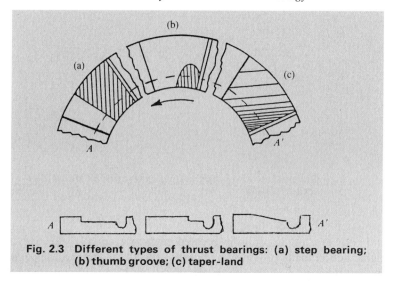

Fig. 2.3   Different types of thrust bearings: (a) step bearing; (b) thumb groove; (c) taper-land

The wedge is one form of restriction; a step is another. It has been realized, for example, that simple thrust collars for motor car engines effectively generate fluid films because shallow slots are cut in them. These were originally provided to make sure that oil got to the load-bearing surface, but in fact they give a series of step restrictions generating fluid pressure. They are best when shallow, i.e., of the order of 20–50 μm. A better form is a thumb-shaped groove that does not extend across the whole width of the bearing, but provides the step restriction at its edge: the oil cannot escape as easily as with straight through grooving.

Figure 2.3 illustrates various types of simple thrust bearings.

THRUST BEARINGS

Typical lubricating film thickness = 0.025 μm

2.2.2   Journal bearings

The oil film wedge in a journal bearing is provided through the clearance. The journal sets itself eccentrically in the clearance space so as to provide a

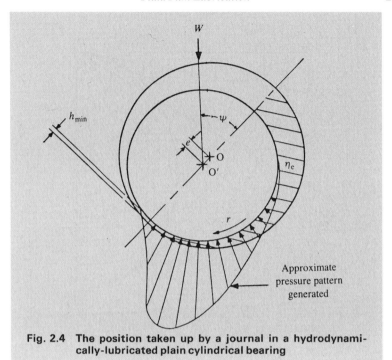

Fig. 2.4 The position taken up by a journal in a hydrodynami-cally-lubricated plain cylindrical bearing

tapering film on the loaded side, generating a pressure film of the shape shown in Fig. 2.4. The displacement of the journal centre from the bearing centre is known as the eccentricity. It is convenient to express this non-dimensionally as the eccentricity ratio (ε), the ratio of eccentricity (e) to the radial clearance of the bearing (c_r).

ECCENTRICITY RATIO

$$\varepsilon = e/c_r$$

When the journal is at the bearing centre, $e = 0$ and hence $\varepsilon = 0$; when it is touching the bearing $e = c_r$ and hence $\varepsilon = 1$. When running, the journal floats on an oil film and is displaced from the bearing centre as shown in

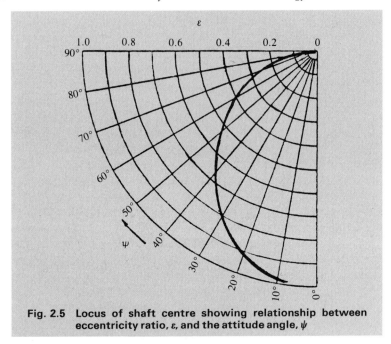

Fig. 2.5   Locus of shaft centre showing relationship between eccentricity ratio, ε, and the attitude angle, ψ

Fig. 2.4. The angle between the load line and the line through the bearing and journal centres is known as the attitude angle, ψ; the line of centres passes through the point of minimum film thickness (h_{min}) in the loaded half of the bearing. The attitude angle varies with the eccentricity ratio as shown in Fig. 2.5.

From a side view, the pressure distribution is parabolic in shape, falling to ambient at the bearing edges (Fig. 2.6(a)). If a circumferential groove is cut into the bearing, the pressure falls to ambient at the groove and the load-carrying capacity is reduced, the bearing being effectively cut into two narrow bearings (Fig. 2.6(b)). Alternatively, if the load to be carried is the same, the two parabolas of pressure will be greatly increased in height, the film thickness at the point of closest approach being greatly reduced.

The clearance in journal bearings is small, of the order of 0.1 percent of the journal diameter, but it must be provided in order to allow a proper

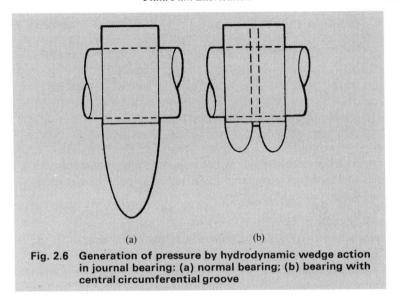

Fig. 2.6 Generation of pressure by hydrodynamic wedge action in journal bearing: (a) normal bearing; (b) bearing with central circumferential groove

film to be generated. As with the thrust pad, the integration of the pressure profile equals the load on the journal and the system is in equilibrium.

It will be seen that the higher the viscosity and the higher the speed of rotation, the better the wedging action and the more concentrically the journal will run. Increase in load has the opposite effect. A qualification should be introduced with respect to speed: as speed increases, naturally the rate of heat generation is increased and this has the effect of reducing the viscosity of the lubricant, so that these two variables are not independent. Increase of width theoretically increases the load-carrying capacity, but in practice there is a rapid diminution in its beneficial effect because of difficulties in maintaining alignment and it is common practice to adopt a width less than the diameter. Increasing clearance beyond a certain minimum value, lowers the load-carrying capacity by increasing the wedging angle and reducing the build-up of pressure.

There is considerable literature dealing with the theoretical treatment of fluid films between moving surfaces. This is very complicated and it is not proposed to refer further to it. However, it is worth deriving the expression for the frictional force generated between two concentric cylinders

JOURNAL BEARNGS

Typical design and operating parameters

b/d	0.5–0.75
c_d/d	1×10^{-3} to 2×10^{-3}
h_{min}	0.025 mm
Peak pressure in the lubricant film	10 N/mm^2

c_d = diametral clearance

separated by a viscous fluid. If the inner cylinder is rotated with a frequency of n then the circumferential frictional force at the surface is

$$F = \eta \frac{\pi dn}{c_r} \pi db$$

where η is the viscosity, d the diameter, c_r the radial clearance, and b the width.

Assuming a load of W, a rather artificial expression for the coefficient of friction can be obtained by dividing the frictional force by the load; this gives

$$\mu = \frac{F}{W} = \pi^2 \frac{d}{c_r} \frac{\eta n}{p}$$

where p is the projected area load, W/bd. The non-dimensional group $\eta n/p$ is an important parameter in hydrodynamic lubrication, analogous to Reynolds Number in fluid dynamics.

Two important variables in the journal bearing are the coefficient of friction and the minimum film thickness. The plot in Fig. 2.7, usually known as a Stribeck diagram, shows the coefficient of friction as a function of the parameter, $\eta n/p$; to the right of the minimum point a fluid film is fully developed. The influence of the main operating variables is evident. More complete non-dimensional parameters are frequently used to take into account the remaining geometric variables of the bearing.

2.3   ELASTO-HYDRODYNAMIC LUBRICATION

The importance of film thickness at the point of closest approach should be emphasized. If the thickness is 0.025 μm it still consists of 100–1,000

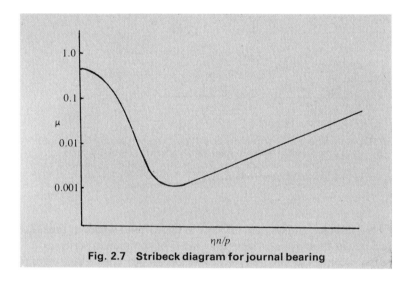

Non-dimensional parameters used with journal bearings

Hersey number $\dfrac{\eta n}{p}$

Sommerfeld number $\dfrac{\eta n}{p}\left(\dfrac{d}{c_d}\right)^2$

Duty parameter $\dfrac{p}{\eta n}\left(\dfrac{c_d}{d}\right)^2\left(\dfrac{d}{b}\right)^2$

Fig. 2.7   Stribeck diagram for journal bearing

molecules of oil and there is ample opportunity for hydrodynamic effects to occur. However, when lubricant films reach this order of thickness the pressure build up is sufficiently great to cause elastic deformation of the surfaces that is significant relative to the film thickness. This is the region of elasto-hydrodynamic lubrication. The development of elasto-hydrodynamic lubrication films means that very much higher loads can be carried on fluid films than would be predicted on simple hydrodynamic theory. Under the very high pressures, not only is there elastic deformation of the surfaces, but the properties of the lubricant can also be affected; in particular, an enormous increase in viscosity may occur so that the liquid behaves more like a solid than a liquid, giving an additional enhancement to load-

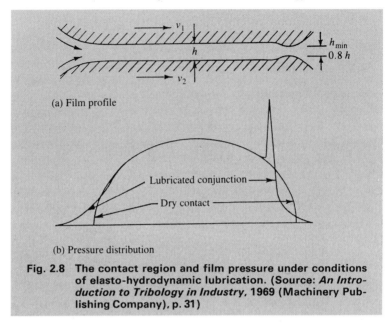

(a) Film profile

(b) Pressure distribution

Fig. 2.8 The contact region and film pressure under conditions of elasto-hydrodynamic lubrication. (Source: *An Introduction to Tribology in Industry*, 1969 (Machinery Publishing Company), p. 31)

carrying capacity. In fact, it is difficult to understand why film breakdown should occur, though that it does is a matter of common experience.

It has been experimentally shown that under elasto-hydrodynamic lubrication an essentially plane contact zone is formed (Fig. 2.8); increasing load increases the length of the contact zone in the sliding direction, but has little effect on the film thickness.

ELASTO-HYDRODYNAMIC LUBRICATION

Typical film conditions

length of contact zone
   in sliding direction   *ca.* 250 μm
film thickness   *ca.* 1 μm
pressure in lubricant film   1500–3000 N/mm^2

Elasto-hydrodynamic lubrication is not just a transition state nor an elegant mathematical conception; it is the operating condition of lubrication in many concentrated contact mechanisms, for example, gears, ball bearings, and rubber lip seals; it also occurs in the aquaplaning of car tyres. The high viscosity of the contact film is of considerable importance in lubricated power transmission equipment such as clutches: it overcomes the difficulty in understanding how the shear stress can be transmitted through a fluid film with very small slip and hence high efficiency.

It is becoming increasingly clear that a number of phenomena that were formerly attributed to boundary lubrication can now be explained in terms of elasto-hydrodynamics. For example, the poor boundary lubricating properties of silicones may be a function of their low pressure–viscosity characteristics rather than, or possibly in addition to, their non-polar nature and their unreactiveness to metal surfaces. Nevertheless, a lubrication regime does exist in which the behaviour cannot be explained in terms of the viscosity or pressure–viscosity characteristics of the lubricant, but rather in terms of physical or chemical reactions between the lubricant and the solid surface. This is the field of true boundary lubrication that was discussed in Chapter 1.

2.4 MIXED LUBRICATION

It is fairly easy to achieve a complete separation of the surfaces by a fluid film. It is, however, difficult to obtain a pure boundary film without any intervention of partial fluid films. Moreover, boundary lubrication is something to be avoided, while fluid film lubrication is something to be fostered, both because of the difference in the coefficient of friction and because having the surfaces separated is the most favourable condition for avoiding wear. As loads are increased, films get thinner and the stage is inevitably reached when some contact occurs. As this is a condition to be approached but not reached for efficient working, it is important to consider the consequences of overloading.

MIXED LUBRICATION

An intermediate lubrication regime when part of the load is carried on fluid hydrodynamic films, part on boundary films.

Consider two slightly rough surfaces just separated by a fluid film and an increasing load that brings them progressively into more and more severe contact. When the first gentle contact occurs, the carpet pile boundary film provides protection. When it becomes a little more severe, the boundary film is mechanically swept off and the frictional heating probably leads to some desorption of the neighbouring film. However, since the contact is a transient one between two opposing high spots, there is an opportunity for the films to be repaired when the surfaces are rewetted by the lubricant and for the heat to be dissipated. Now imagine the contact to be still more severe so that not only is that boundary film displaced but oxide-to-oxide contact occurs; a particle may be detached and considerable heat generated locally. If it is still only an odd high-spot contact, the heat will be dispersed by the lubricant flowing over the surfaces and boundary films formed on the newly exposed surfaces. By a succession of such events the surfaces will be made appreciably more conforming and this will favour fluid film formation. The situation is thus better than it was before. This is the mechanism of 'running-in', i.e., the improvement of surfaces by removing high spots to make them better able to maintain a full fluid film at higher loads.

> **RUNNING-IN**
>
> The removal of high spots in the contacting surfaces by wear or plastic deformation under controlled conditions of running giving improved conformability and reduced risk of film breakdown during normal operation.

Figure 2.9 shows surface traces of a virgin machined surface and the same surface after running-in. Sometimes instead of high spots being removed, they are plastically deformed, but the effect is the same, namely to make the surfaces more conforming and enabling higher loads to be carried without film breakdown. This is illustrated in Fig. 2.10 which shows the first and third traces of Fig. 2.9 superimposed on each other at a separation of 5 μm (lubricant film thickness is considered to be the distance between the hypothetical centre lines of the two surfaces referred to in Chapter 1 when discussing surface roughness). This figure shows the greater effective separation of the run-in surfaces

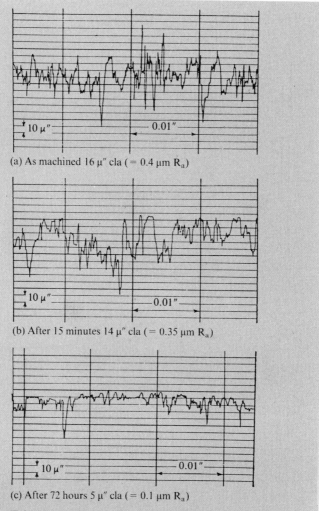

(a) As machined 16 μ'' cla ($= 0.4\ \mu$m R_a)

(b) After 15 minutes 14 μ'' cla ($= 0.35\ \mu$m R_a)

(c) After 72 hours 5 μ'' cla ($= 0.1\ \mu$m R_a)

Fig. 2.9 Talysurf traces of mild steel surface showing the effect of running-in

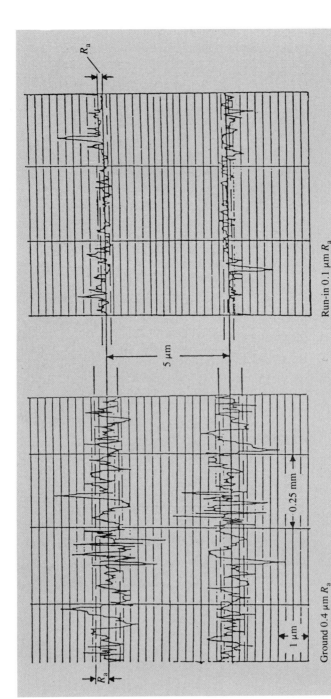

Fig. 2.10   Effect of run-in on actual face separation

or, alternatively, how increased load could be applied to the run-in surfaces before they approach the separation condition with the original machined surfaces.

If the load is increased still further, not only will the heat generated be greater, but contacts will be more frequent. There is likely, therefore, to be a reduction in the effective viscosity of the lubricant and it will be less able to sustain its share of load-carrying. There is now a deteriorating condition leading to still more and more severe solid contacts, and finally to scuffing and seizure. Scuffing is the tearing of the surfaces with destruction of their original form, seizure is the condition when the increase of friction leads to the mechanism stalling.

Normally, the usual action is to keep the load and other variables such that this critical condition is not reached; but there are sometimes good reasons for approaching it and there are lubricants that, because of their chemical activity, help to cope with the local heat condition better than plain mineral oils. These are the so-called 'extreme-pressure' lubricants referred to in the previous chapter.

The removal of particles from the surfaces and the further removal of material when these particles get trapped between surfaces constitutes wear. When it is mild and associated with running-in it is beneficial, but normally there is deterioration. The most effective way of stopping wear is to create fluid film conditions and to prevent extraneous particles from getting between the surfaces.

We can now see the effect of surface finish on lubrication. This is readily visualized by the specific film thickness, λ, the ratio of the lubricant film thickness, h_{min}, to the combined roughness of the two contacting surfaces, σ.

SPECIFIC FILM THICKNESS

$$\lambda = \frac{h_{min}}{\sigma}$$

h_{min} = film thickness between hypothetically flat surfaces

σ = composite surface roughness = $(\sigma_1^2 + \sigma_2^2)^{0.5}$

$\sigma_1 = R_q$ of surface 1, $\sigma_2 = R_q$ of surface 2

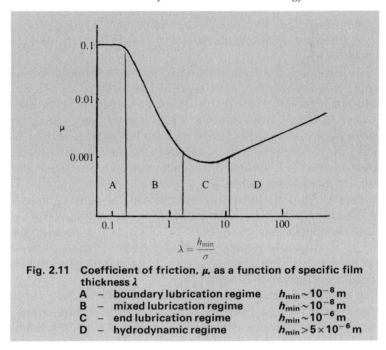

Fig. 2.11   Coefficient of friction, μ, as a function of specific film thickness λ

A	– boundary lubrication regime	$h_{min} \sim 10^{-8}$ m
B	– mixed lubrication regime	$h_{min} \sim 10^{-8}$ m
C	– end lubrication regime	$h_{min} \sim 10^{-6}$ m
D	– hydrodynamic regime	$h_{min} > 5 \times 10^{-6}$ m

Figure 2.11 is a plot of coefficient of friction against specific film thickness. This closely resembles the Stribeck diagram shown in Fig. 2.7, and shows the transition from hydrodynamic lubrication on the right, when the film thickness is about ten times greater than the combined surface roughnesses, to elasto-hydrodynamic lubrication when the film thickness is of the same order as the surface roughness, then to mixed and finally boundary lubrication as the surfaces come into solid contact.

2.5   SQUEEZE FILMS

So far hydrodynamic lubricating films generated by relative motion parallel to the surface have been considered. Equally however, if one surface approaches another, then pressure is normally also generated in any fluid

DYNAMIC PROPERTIES OF LUBRICANT FILMS

Oil films subject to dynamic loads exhibit both spring and damping effects.
This has two consequences:

(a) damping characteristics – give *squeeze-film lubrica-tion* – the ability to carry alternating and transient loads

(b) spring characteristics – possibility of resonance with other springs in the machine system giving rise to *oil-film instabilities*

between them in order for it to be squeezed out sideways. It does this easily whilst the surfaces are far apart, but as they move closer and closer together the effect is to slow down the rate of approach. With very thin films the difficulty of squeezing out the fluid increases very sharply. In fact, the rate of approach is inversely proportional to h^2, where h is the distance between the surfaces. If the film is of the order of that in a hydro-dynamically-lubricated bearing, the time can be very long indeed. It is a recognized phenomenon that smooth, conforming surfaces covered with a film of oil are not easily brought into solid contact.

If the load bringing the surfaces together is alternating, it is probable that with these long times to establish contact, the load will reverse before the separating film breaks down and the surfaces actually touch. Conversely, when the load reverses, the need to suck in lubricant slows down the rate of separation. This is the regime of squeeze-film lubrication. Unlike normal hydrodynamic lubrication, the separation of the surfaces under load can only be maintained for a limited time; however, this is very effective in sustaining reciprocating and shock loads, since it behaves handsomely as a cushion. The film thus has both damping and spring characteristics, a variable-rate spring being soft when the surfaces are well apart and stiff when they get very close together. In a journal bearing, this means soft when the journal is nearly concentric with the bearing, stiff when the eccentricity is high.

It follows as a penalty that such films can be set in resonance and that they can combine with other springs in the system to cause troublesome vibrations. The subject of bearing vibration is a complex one, but as it can lead to failure, particularly in high-speed machines, it is worth briefly mentioning the different phenomena.

VIBRATIONS IN JOURNAL BEARINGS

Synchronous whirl – response to out-of-balance forces
– frequency equals rotational frequency

Half-speed whirl – vibration induced by dynamic characteristics of oil film
– frequency slightly below half rotational speed

Resonant whip – condition when half-speed whirl locks on to system critical
– frequency that of system critical

Synchronous whirl

Out-of-balance mass on the rotor causes the journal centre to orbit round its steady load equilibrium position in the bearing. This orbiting results in vibration of the bearing housing with a frequency equal to the rotational frequency of the rotor. The amplitude of vibration is controlled by the damping properties of the bearing oil film and the stiffness of the bearing mounting. However, if the vibration frequency coincides with a natural frequency of the rotor a resonant condition can build up resulting in a high level of vibration; this resonant vibration condition is known as a synchronous whirl. Provided such critical speeds are passed through quickly the damping action of the oil film slows the rate of build up of the resonance and there is no problem. However, it should be noted that the natural frequency is affected by the bearing oil film – it is reduced compared to that on stiff supports, and for this reason it is not easy to predict the actual critical speeds of rotors.

Half-speed whirl

Half-speed whirl is a bearing instability that arises from the dynamic characteristics of bearing oil films. It can be seen from Figs 2.4 and 2.5 that, if a small perturbing force is added to the load, not only does the eccentricity increase, but the attitude angle also changes (decreases). When the perturbing force is removed, the spring force in the oil film acts to restore the journal to its original position; however, because of the change in attitude angle, the restoring force is not collinear with the original disturbing force and the net effect is that the journal tends to orbit round its original centre, marked as O′ in Fig. 2.4. This orbit is reinforced by a repeated perturbing force. The whirling shaft now becomes a pump with the whirl frequency set at the speed at which the shaft can pump the oil round the bearing. This is most easily visualized for the simple case of the journal whirling round the centre of the bearing with the eccentricity ratio close to zero. The average lubricant velocity in these circumstances is half the peripheral speed. Hence the greatest frequency at which a pressure pattern can progress round the bearing is half the rotating speed, falling below half with increasing eccentricity. (This applies to the case of a simple cylindrical bearing. It is possible to establish 'half-speed' whirls in non-cylindrical bearings at frequencies greater than half-speed.)

The pressure pattern set up corresponds to a load at approximately half shaft frequency. Problems arise because no hydrodynamic lift is generated for such a load. Figure 2.12(a) shows a bearing with a load rotating at half frequency. This is physically the same as holding the load line constant, with the journal and bearing each rotating at half frequency in opposite directions, Fig. 2.12(b), and from this it can be seen that there is no net flow into the load zone and hence no load-carrying capacity. If a shaft is rotated at twice its first natural frequency, then any precessional motion can create a load rotating at half shaft frequency and, as there is no hydrodynamic pressure generated to carry such a load, a resonance can develop. The actual situation in a bearing is more difficult to predict as the effect of the bearing oil film with its spring characteristics, which depend on the eccentricity ratio, is to reduce the natural frequency of the shaft. However, this does give a simplified picture of the origin of half-frequency whirl, though for the reason given above it does not occur at exactly half the rotational frequency, but somewhat less, usually in the range 0.4–0.49.

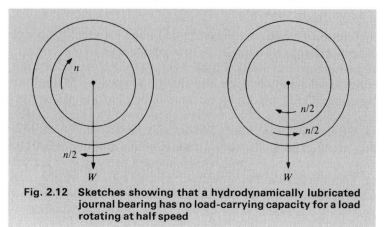

Fig. 2.12   Sketches showing that a hydrodynamically lubricated journal bearing has no load-carrying capacity for a load rotating at half speed

Resonant whip

With flexible shafts vibration frequency can lock on to the natural frequency and does not die away with increasing shaft speed. This is the condition of resonant whip.

Both half-frequency whirl and resonant whip can give rise to severe vibrations of high amplitude and, though squeeze-film effects usually prevent breakdown of the oil film, the high pressure fluctuations in the oil film can cause fatigue failure of the bearing in both loaded and nominally unloaded halves. These oil-film instabilities are most severe when running near or above rotor natural frequencies and special designs may be necessary to allow successful operation.

2.6   GAS FILM LUBRICATION

So far this discussion on fluid film lubrication has been with reference to liquid films. This is, of course, by no means the full story and gas can equally well be used. The use of self-generated gas lubricating films is usually referred to as aerodynamic lubrication, though of course it is not restricted to air. As far as lubrication is concerned gases differ from liquids in four important aspects.

- The viscosity of gases is low – about a tenth that of the most mobile liquids with little variation between different gases (a factor of two or three at most).

- The viscosity of gases is almost independent of temperature and pressure, at least at the pressures obtaining in bearings.

- Gases have no boundary lubricating properties.

- Gases are highly compressible when compared with liquids.

It is worth indicating briefly the physical significance of these differences as far as fluid-lubricating bearings are concerned without going into the mathematics of gas-lubricated bearings, which are more complicated than those of liquid-lubricated ones.

It is important to realize that there is a very significant difference between the load-carrying capacity of a bearing under boundary and hydrodynamically-lubricated conditions: the load-carrying capacity of a bearing started under a static load is usually determined by the former unless some form of hydrostatic jacking is used to cover the start-up. With oil-lubricated bearings a typical figure may be taken as $3.5 \, \text{N/mm}^2$, though once a fluid film has been generated it may be possible to carry peak dynamic loads of more than $70 \, \text{N/mm}^2$. In contrast, because of the lack of boundary-lubricating properties, the maximum start-up loading with gas-lubricated bearings has to be restricted to $0.05–0.1 \, \text{N/mm}^2$ and the bearing materials have to be selected so that they are not damaged on starting and stopping.

A consequence of the lower viscosity of gases is that for the same speed of rotation the film thickness will be lower than for an oil-lubricated bearing and this requires better surface finish and greater care in alignment to prevent breakdown at the edges of the bearing. This problem is rendered more acute by the fact that it is desirable to make the width : diameter ratio of gas bearings higher than for oil-lubricated bearings in order to compensate for the reduced pressure generation in the lubrication film. This problem can be overcome to some extent by mounting the bearing on a flexible diaphragm so that it is self-aligning.

The low viscosity of gases on the other hand means that the viscous friction losses are reduced. This can be a very real advantage in high-speed bearings. The fact that the viscosity of gases does not decrease with temperature is also of value in this situation.

In addition to the fact that high compressibility tends to result in decreased film thickness, unless the gas is fed into the bearing at high pressure, it also means that gases have less damping capacity than oils and hence that gas-lubricated bearings have a greater proneness to lubricant-film-induced instabilities than oil-lubricated ones.

Because of these very considerable difficulties, gas-lubricated bearings tend to find application only in special circumstances. One example is the high-speed, turbine-driven dental drill; by using the air for the turbine as a lubricant for the bearings, speeds of 10,000 rev/s can be achieved and the need for a lubricant that is acceptable to the patient is eliminated. In the dental drill the low load-carrying capacity is a positive advantage; if the dentist presses too hard the film breaks down and the increased friction stalls the turbine!

A further attraction of the gas bearing is that it eliminates the risk of oil contamination that can be a problem in certain industrial processes, the gas that is handled being used for lubrication. An example of this is in carbon dioxide circulators used on nuclear energy plant, where by using the circulated gas for lubrication the risk of contamination of nuclear reactors can be eliminated.

Gas-lubricated bearings call for high standards of engineering manufacture and fitting, to say nothing of high standards of cleanliness, but they present real opportunities for effective exploitation in suitable circumstances.

CHAPTER 3

Lubricants

Viscosity is the most important property of a lubricant, though some *boundary lubricating properties* are required for start-up in hydrodynamically-lubricated mechanisms.

Mineral oils, obtained from petroleum, form the basis of the majority of industrial lubricants, though small amounts of *additives* may be incorporated to enhance specific properties:

anti-oxidants	–	to extend lubricant life
corrosion inhibitors	–	to protect steel parts against corrosion
load-carrying additives	–	to prevent damage of heavily-loaded contacts in event of lubricant film breakdown.

Greases i.e., thickened oils that do not flow under their own weight, and *solid lubricants* extend the application of liquid lubricants.

A lubricant has two functions: (a) the elimination of damage to the sliding surfaces, and (b) the reduction of friction between them. Let us look at these requirements and at some other desirable properties in lubricants before considering the materials that are used.

3.1 PROPERTIES REQUIRED IN LUBRICANTS

3.1.1 Viscosity

The primary objective in lubrication is to separate the surfaces completely by a fluid film. This is the most favourable condition for avoiding damage and also allows friction to be at a minimum. The plot of μ against $\eta n/p$ (Fig. 2.7) shows that viscosity is the lubricant's only significant property as far as fluid film formation is concerned. In order to operate with minimum friction loss the viscosity has to be adjusted to suit the operating conditions of load and speed. For high speeds and low loads, gases can be used, but for the majority of industrial applications it is normal to use liquid lubricants.

Because of the importance of viscosity in lubrication, it is perhaps worth defining it and describing some of the units in which it is expressed. When relative motion takes place between adjacent layers of a fluid, internal shear stresses arise to oppose the motion. This resistance to flow is known as viscosity. Fluids in which the shear stress, s, is directly proportional to the rate of shear, du/dy, are known as Newtonian fluids, the constant of proportionality being the dynamic viscosity (sometimes called the absolute viscosity).

DYNAMIC VISCOSITY

$$q = \eta \frac{du}{dy}$$

The SI unit of dynamic viscosity is N s/m^2; however, the unit frequently used in lubrication is the cgs unit, the Poise, P, or more commonly the centiPoise, cP, where

$1\ cP = 0.01\ P$

$\quad\quad = 10^{-3}\ N\ s/m^2$

In practice, since the viscosity of lubricants is commonly measured in a 'U' tube viscometer, where the lubricant flows under the action of gravity, giving the kinematic viscosity, lubricant manufacturers tend to quote kinematic viscosity rather than dynamic viscosity. The SI unit of kinematic viscosity is m^2/s, but again it is the cgs unit, the Stokes, St, or more commonly the centiStokes, cSt, that is used in practice.

KINEMATIC VISCOSITY

$$1\ cSt = 1\frac{cP}{\rho} = 10^{-6}\ m^2/s$$

where ρ = density in g/ml (10^{-3} kg/m^3)

Gases have characteristically low dynamic viscosities and there is little difference in viscosity between different gases – a factor of three at most. The viscosity of a gas increases slightly with increasing temperature. In contrast, the viscosities of different liquids vary widely and the viscosity of any particular liquid markedly decreases with temperature, the rate of decrease being more marked the greater the viscosity (Fig. 3.1). The viscosity of the most mobile liquids at room temperature is approximately an order of magnitude greater than that of a gas, and the different liquids available cover a range over four orders of magnitude. Liquids are by far the most important in industrial practice and further discussion will be concentrated on them.

3.1.2 Boundary-lubricating properties

Even with bearings designed to operate with fluid films, some solid contact will occur during starting and stopping. Unless the unit loading is very low the lubricant must possess adequate boundary lubricating properties to prevent surface damage in these situations. In fact, it is normally the conditions at start-up that determine the permissible loading. With gases that are completely deficient in boundary-lubricating properties, static loadings have to be limited to about 0.05–0.1 MPa (N/mm^2); with liquid lubricants the limiting figure is about 3.5 MPa.

3.1.3 Other requirements

A number of other features are desirable in a lubricant. In particular, its properties should not change with exposure to the operating conditions. This means good mechanical stability, a wide liquid range, good thermal stability, and resistance to oxidation. Most industrial machines are made of iron or steel and it is important that the lubricant should protect finely machined parts against corrosion. Finally, lubricants should not be toxic or harmful to operators and, if possible, cheap and readily available.

3.2 MINERAL LUBRICATING OILS

A wide range of liquids could be used as lubricants, but the above requirements are met most admirably by mineral hydrocarbon oils, obtained from the heavier fractions produced in the refining of petroleum.

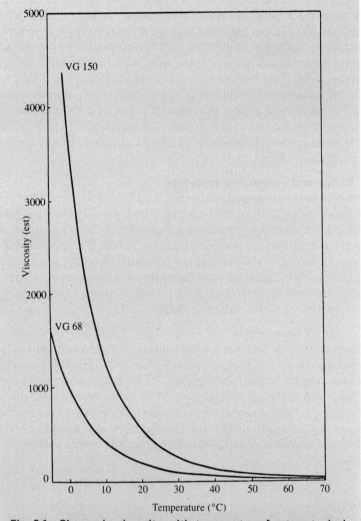

Fig. 3.1   Change in viscosity with temperature for two typical mineral oil lubricants

Petroleum is a complex mixture of hydrocarbons, chemical compounds of carbon, and hydrogen, with widely differing molecular type and size. While it is not necessary for the engineer to go deeply into the chemistry of mineral oils, some basic understanding can be useful in discussions with lubricant suppliers and in investigating tribological failures.

There are three basic hydrocarbon types: paraffins, naphthenes, and aromatics, Fig. 3.2. Paraffins consist of chains or branched chains of carbon atoms with the carbon atoms fully saturated, that is each carbon atom is connected to its neighbouring atoms by single valency bonds (represented by a single line in Fig. 3.2). Naphthenes consist of ring structures of carbon, again with the carbon atoms fully saturated. Aromatic hydrocarbons again have ring structures, but in this case the carbon atoms are connected by alternating single and double (unsaturated) valency bonds (represented by double lines in Fig. 3.2).

In mineral lubricating oils the hydrocarbon compounds are much more complex than those illustrated in Fig. 3.2, since they are mixtures of the

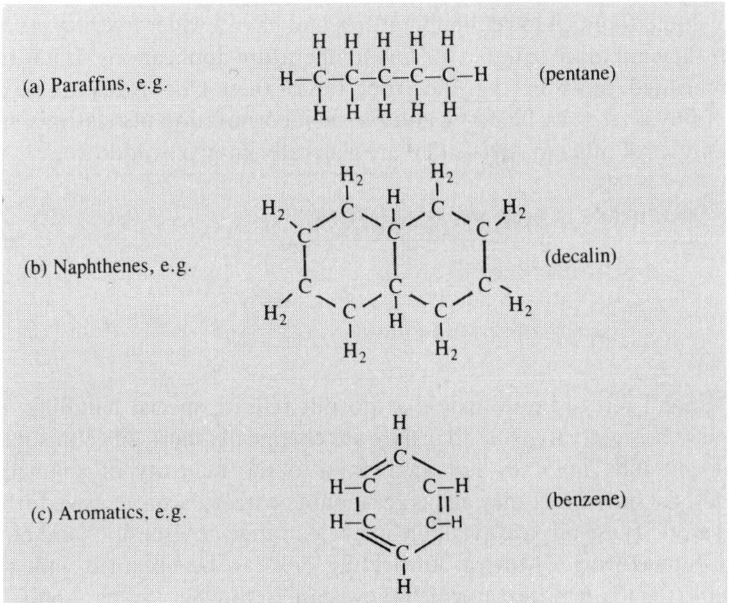

Fig. 3.2   Basic hydrocarbon types

basic hydrocarbon types and typically contain 12–60 carbon atoms. The lubricating oils produced depend on the crude petroleum used and the nature of the refining process. They are described as 'paraffinic' or 'naphthenic', depending on the major characteristics of the hydrocarbon molecules present: oils intermediate in character are described as 'mixed base'. All mineral lubricating oils contain about 15–20 percent aromatic-type molecules.

The properties depend to some extent on the basic hydrocarbon type. In general, the rate of change of viscosity with temperature is lower with paraffinic-type oils than with naphthenic ones. Paraffinic oils tend to be more resistant to oxidation than naphthenic ones; however, when they do break down at high temperature, as for example in the cylinders of engines or reciprocating compressors, they tend to form a harder 'carbon' (although described as such the deposit is not actually carbon, but a highly oxidized and polymerized degradation product) that is more difficult to remove. At low temperatures paraffinic oils tend to precipitate wax, and solidify over rather a narrow temperature range, whereas naphthenic oils stay liquid to much lower temperatures and have a wider freezing range so that they are more suitable for low temperature applications. It has to be appreciated, however, that the properties of these basic types can be considerably altered by blending and by the incorporation of relatively small quantities of other materials that are generally known as additives.

PROPERTIES OF MINERAL OILS

   – Chemically inert
   – Available in wide range of viscosity
   – Wide temperature range

Mineral oils are non-toxic and do not require special handling techniques. It is a great virtue that they are chemically inert, since this means that not only are they not aggressive to the majority of engineering materials, but in fact they act as reasonable corrosion protectives for iron and steel. These oils are available in a wide range of viscosities and have a wide liquid range. Mineral lubricating oils can be obtained with pour points (the temperature at which they cease to flow) as low as $-60°C$ and they are stable in the absence of oxygen to over $300°C$.

3.2.1 Viscosity characteristics

One of the most important features of mineral oils is that they can be obtained in a continuous spectrum of viscosities ranging from liquids as mobile as water to ones that will scarcely flow at room temperature. To differentiate between the viscosities of different lubricants it is necessary to define the viscosity at some prescribed temperature. The agreed temperature for industrial lubricating oils is 40°C. The viscosity in cSt at 40°C is known as the viscosity grade (VG).

In the interest of rationalization, the viscosity range has been divided into a number of discrete viscosity grades as shown in Table 3.1; this is based on a logarithmic series with six grades in each decade of viscosity extending from 2 to 1,500 cSt. For manufacturing purposes a tolerance of ± 10 percent about the grade mid-point value is permitted. It will be seen that there are gaps in the classification. However, this is not of great significance in selecting oils for practical applications, largely because of the marked way the viscosity changes with temperature.

Table 3.1 ISO viscosity classification for industrial liquid lubricants

ISO viscosity grade	Mid-point kinematic viscosity cSt at 40°C	Kinematic viscosity limits cSt at 40°C	
		Minimum	Maximum
ISO VG 2	2.2	1.98	2.42
ISO VG 3	3.2	2.88	3.52
ISO VG 5	4.6	4.14	5.06
ISO VG 7	6.8	6.12	7.48
ISO VG 10	10	9.00	11.0
ISO VG 15	15	13.5	16.5
ISO VG 22	22	19.8	24.2
ISO VG 32	32	28.8	35.2
ISO VG 46	46	41.4	50.6
ISO VG 68	68	61.2	74.8
ISO VG 100	100	90.0	110
ISO VG 150	150	105	165
ISO VG 220	220	198	242
ISO VG 320	320	288	352
ISO VG 460	460	414	506
ISO VG 680	680	612	748
ISO VG 1000	1000	900	1100
ISO VG 1500	1500	1350	1650

It has to be recognized that the viscosity of all types of oil changes rapidly with temperature. This causes great difficulties in calculating the exact operating conditions in bearings, but allows for considerable simplification in oil grade selection. The operating temperature of a bearing is normally the most critical factor. It is determined by the balance between the heat generated by viscous friction in the bearing, a function of the viscosity, and the heat removed by cooling. If the viscosity grade of oil is increased, this results in an increase in the coefficient of friction (Fig. 2.7) and a corresponding increase in the temperature, with a consequent reduction in the oil viscosity. The converse occurs if the viscosity grade is reduced. However, because of the rapid change of viscosity with temperature, the new equilibrium is established at a temperature only a few degrees away from the previous value.

Hydrodynamically-lubricated bearings have remarkable self-stabilizing properties and it is practical experience that bearings show a wide tolerance to changes in viscosity grade with comparatively little change in operating parameters.

This characteristic enables the range of viscosity requirements for lubrication to be met by a quite restricted range of viscosity grades. Figure 3.3 shows the viscosity–temperature characteristics of a typical series of oils covered by the ISO/BS classification. The insensitivity of lubricated mechanisms to the viscosity grade of the oil used allows for a considerable rationalization of the lubricants stocked. In any one plant or factory it should be possible to meet all requirements with alternate members of the classification.

3.2.2 Viscosity index

The other viscosity property that is of some significance is the viscosity's rate of change with temperature. This is normally expressed by the viscosity index (VI), which is an arbitrary number originally going from 0 to 100, (now extended to 140), based on the viscosities at $40°C$ and $100°C$. The higher the VI the lower the rate of change of viscosity with temperature. There are real differences between the viscosity temperature characteristics of paraffinic and naphthenic oils; naphthenic oils tend to have low VIs,

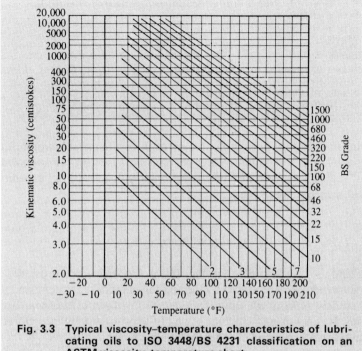

Fig. 3.3 Typical viscosity–temperature characteristics of lubricating oils to ISO 3448/BS 4231 classification on an ASTM viscosity–temperature chart

whereas paraffinic ones have high VIs. In practice, however, VI has little significance in industrial lubrication applications.

3.2.3 Boundary-lubrication properties

The boundary-lubricating properties of mineral lubricating oils are adequate to cover normal operating conditions. Typical values for two viscosities of plain mineral oil are given in Table 3.2 for a steel on steel contact.

It will be noted that the coefficient of friction begins to rise at about 140°C, more so with the lower viscosity oil. This corresponds to the melting point of the adsorbed soap film. This is not, however, a concern in most mechanisms. Both white-metal lined plain bearings and rolling bearings should not be operated at temperatures above 140°C. White metal

Table 3.2   Boundary-lubricating properties of mineral oils

Lubricant	Temperature	Coefficient of friction (steel on steel)
None	0–200°C	0.60
Low viscosity oil	20°C	0.12
(32 cSt at 40°C)	140°C	0.40
High viscosity oil	20°C	0.12
(1000 cSt at 40°C)	140°C	0.14

bearings because of the loss of strength of the white metal limiting its use to about 120°C maximum, rolling bearings because they are annealed at about this temperature and will suffer reduced load-carrying capacity if used at higher temperatures.

3.3   LUBRICATING-OIL ADDITIVES

It should be emphasized that the above properties are possessed by straight mineral oils produced by the refining of crude petroleum, and that such straight mineral oils can satisfy the vast majority of industrial applications. It is possible to enhance some of the properties by incorporating relatively small amounts of other materials, known as additives. It should not be thought, however, that additive-containing oils are necessarily better than straight oils; they are different and can give real advantages in certain situations.

LUBRICANT ADDITIVES

Additives are in most cases more reactive than the base oil, and this can at times cause problems. The discriminating user should, therefore, know something of their advantages and disadvantages.

The additive content of industrial lubricating oils is low, normally less than 1 percent. IC engine oils contain much higher levels, in some cases up to 15 percent. Such oils are not required for the lubrication of industrial machines.

MAIN ADDITIVE TYPES IN INDUSTRIAL LUBRICANTS

- Anti-oxidants
- Corrosion inhibitors
- Load-carrying additives: anti-wear
  extreme-pressure

3.3.1 Anti-oxidants

Mineral oils oxidize at elevated temperatures in contact with air, the rate of oxidation being a function not only of the temperature but the nature of the contact with air. Figure 3.4 shows the effect of temperature on the life of mineral oils as determined by oxidation stability; the lines have been deliberately thickened to show the spread of values between different types of oils. As a rough guide it can be said that oxidation below 40°C is so slow that it can be ignored. At 60°C, however, the life may be reduced to about a year and above this temperature it reduces rapidly. It should be noted that these lives refer to intimate contact with air. In a bearing where access of air is limited, and in any case where the time of exposure is short, much higher temperatures will not cause serious degradation.

The increased stability of oils containing anti-oxidants is also shown in Fig. 3.4. The anti-oxidants that are incorporated in industrial mineral oil lubricants, usually at concentrations of only 0.1–0.5 percent, are effective to temperatures about 80°C; above this they either volatilize or break down. The aim is to enhance the life of the oil in the reservoir or sump, not in the bearing. Anti-oxidants are particularly effective for large circulating oil systems, where with normal top-up the life of the oil containing them can equal that of the machine.

3.3.2 Corrosion inhibitors

It has already been mentioned that mineral lubricating oils possess reasonably good corrosion-preventing properties for ferrous surfaces. This depends on the presence of small amounts of polar materials that form by the oxidation of the oil. These materials do not form if an anti-oxidant is added to the oil and the deficiency has to be made up by incorporating an anti-corrosion additive, normally at about 500 ppm. Circulating oils containing both anti-oxidation and anti-corrosion additives are frequently referred to as double-inhibited oils.

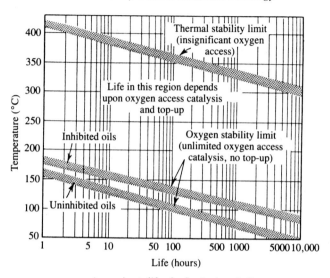

Approximate life of refined mineral oils

**Fig. 3.4 Thermal and oxidation stability of mineral hydrocarbon lubricants.
(Source: T. I. Fowle, *Tribology Handbook*, First edition, 1973
(Mechanical Engineering Publications, London))**

3.3.3 Anti-wear and extreme pressure (ep) additives

Under extreme conditions of frictional contact the boundary-lubricating
film can be destroyed, normally desorbed through high local temperature. A range of additives based on oil-soluble compounds of sulphur,
phosphorus, and chlorine can be incorporated in the oil to prevent
damage. Such materials act by reacting chemically with the surface to
inhibit adhesion between the surfaces that would lead to scuffing and
seizure. The mechanism of action is thus a sort of sacrificial corrosion
designed to inhibit more severe damage. They are potentially chemi-
cally-reactive materials that can cause problems in some circumstances
and should thus only be used where they are needed. Table 3.3 outlines
the loading conditions where anti-wear and ep additives can give real
benefit.

Table 3.3 Heavily-loaded contacts in industrial machines where load-carrying additives should be used

Gears	Spur and helical	Lloyd's 'K' factor* $> 3.5\,N/mm^2$ $(500\,lbf/in^2)$
	Hypoid	All
Sliding-vane pumps		Discharge pressure > 140 bar
Rolling bearings	Ball bearings	$C/P < 4$
	Roller bearings	$C/P < 6.5$
	or	
	all	Specific film thickness < 1.2

* Lloyd's 'K' factor is a practical criterion used to express the stress condition between gear teeth. It is given by the expression

$$K = \frac{F_t\,(r+1)}{d\ \ r}$$

where F_t = tangential force/unit width of tooth face. N/mm
        d = pinion pitch-circle diameter
        r = gear ratio.
        c = basic dynamic capacity of rolling bearing
        p = equivalent radial load, rolling bearing

3.3.4 Anti-foam additives

The build-up of foam on the surface of the lubricant in the reservoir can be a problem. Very small amounts of certain long-chain polymers, particularly polymethylsiloxanes, are remarkably effective in breaking down foam and have been incorporated in lubricants as anti-foam additives.

However, it has been found that such additives can have an adverse effect by slowing down the separation of air bubbles from the oil, and it has been found preferable to obtain good air separation and foam breakdown properties by suitable refining of the base oil rather than by the use of additives.

3.3.5 Detergent additives

Detergents, or more accurately dispersant additives, are incorporated in engine oils. Strictly they have nothing to do with lubrication as such: their function is to keep in suspension any carbonaceous particles from the fuel that might otherwise collect on the piston crown and in the piston ring area, gradually forming deposits that would interfere with the functioning of the engine. They have little application in industrial lubricants.

3.4 GREASE

Greases are oils that are thickened with solids to give a
semi-fluid product.

In general, the use of greases is restricted to lower speeds since they do not
flow readily and cannot be used to conduct away the heat that is generated
at high speeds. Conversely, the resistance to flow means that there is no
need for the elaborate sealing arrangements that are required in oil-
lubricated systems to keep the lubricant in place. Greases are used particularly for rolling element bearings, where, not only is it easy to retain the
lubricant, but the grease can also act as its own seal, thereby keeping dirt
out. Semi-fluid greases are used in hand tools, again taking advantage of
easier containment.

The commonest thickening agents are metal soaps. The properties of
the grease depend on the viscosity of the base oil, the soap type, and quantity. Table 3.4 shows some characteristics of the most important grease
types.

A significant characteristic of greases is the drop point, i.e., the temperature at which a grease melts under certain prescribed conditions. This
gives some guidance on the maximum temperature to which a grease can
be used (when a grease becomes liquid it loses any advantage over an oil),
but in practice the upper limit is determined by a combination of thermal
and mechanical stability. Practical limits for short and long term operation are suggested in Table 3.4.

Table 3.4 Common grease types

Thickener	Calcium (lime)	Lithium	Lithium complex	Inorganic (clay)
Drop point °C	100	180	230	None
Temperature range °C:				
short term	0 to 60	−20 to 140	−30 to 175	10 to 200
long term	0 to 60	−20 to 70	−30 to 100	−
Corrosion-preventing properties	Very good	Good	Good	Fair
Relative cost	1	1.5	3	2

Calcium (lime) greases are the cheapest, but are limited in temperature range. They are stabilized by water and cannot be used either below 0°C, where the water would freeze, or above 60°C, where it would evaporate off. They have excellent corrosion-preventing properties in their own right and are often incorporated in corrosion preventives.

Lithium and lithium complex greases have higher temperature stability and are used particularly in rolling bearings. They do not have the same corrosion-preventing properties as calcium greases and corrosion inhibitors are normally incorporated.

Clay greases have no true drop point. Their temperature limit is set by the oxidation stability of the base oil. They are, however, less mechanically stable than soap base greases and are best suited to short-term applications; for example, kiln-car bearings where temperatures are high but exposure time short.

The greater the proportion of soap the stiffer the grease. A consistency classification has been produced by the National Lubrication Grease Institute of America (NLGI); this contains eight grades, 00, 0, 1–6; stiffness increases with the NLGI grade number. The 00 and 0 grades are semi-fluid products that flow under their own weight. The choice of consistency grade depends on the application.

NLGI GREASE CLASSIFICATION

Grade 00 ⎫	Semi-fluid –
⎬	mainly used in hand tools
0 ⎭	and small gearboxes
1	Soft – suitable for pumped systems
2	Rolling bearings
3	Rolling bearings, especially vertically mounted
4 ⎫	Little industrial use
5 ⎭	
6	Block greases

While both Nos 2 and 3 are used for rolling bearings, No. 3 greases are preferred for vertically-mounted bearings to reduce the risk of the grease continually slumping into the bearing. Block greases are used in a cavity bearing where there are negligible arrangements to retain the grease; in a cavity bearing, a 'block grease' of No. 6 consistency would be used.

The base oil used is normally a medium viscosity grade, VG150 or VG220, and the soap content is about 10 percent. It is also possible to make greases from high viscosity oils and in this case only about 3 percent soap is required to obtain the same consistency. Greases of this type are extremely tacky and very resistant to being squeezed out. They are, however, very difficult to pump and are not even easy to dispense with a grease gun. They find application in grease-packed flexible couplings.

3.5 SYNTHETIC LUBRICANTS

Although mineral oils cover the vast majority of industrial applications, they have some limitations and liquid lubrication can be extended by the use of other materials.

Mineral oils are widely used as power transmission fluids in hydraulic systems, where advantage is taken of their good lubricating and corrosion-preventing properties. However, since they are flammable, they can be a hazard in hydraulic systems used in fire-risk areas, e.g., in coal mines and foundries; a fine spray of hydraulic fluid leaking from a high-pressure system presents a much greater risk of fire than drips from a normal lubrication system. Materials with greater resistance to fire can be used in such applications, either by incorporating water, e.g., water-in-oil emulsions, polyalkyleneglycol water mixtures, or materials inherently less flammable, e.g., phosphate esters. It should be noted that, while giving reduced fire risk, these materials can still burn under certain circumstances. They are best described as 'fire-resistant fluids'.

The only genuine non-flammable lubricants are based on the fully fluorinated compounds. These materials are completely inert to oxygen and can be used on machines handling liquid oxygen or as the base for greases for very high temperature applications (ca 300°C), though they are extremely expensive, i.e., 300–400 times that of mineral oil lubricants.

Other synthetic lubricants are available where such extreme temperature resistance is not required. Polyalphaolefines (PAOs), which consist of synthesized hydrocarbon molecules similar to those in normal mineral oils, but specially selected for stability, are used for high-temperature

SYNTHETIC LUBRICANTS

Polyalphaolefines (PAOs)	– synthetic hydrocarbons more stable than mineral oil – suitable for high temperatures
Polyalkaleneglycols (PAGs)	– excellent boundary-lubricating properties
	– solutions of water-soluble grades in water have good fire resistance
Diesters	– low-viscosity oils with wider temperature range than mineral oils
Phosphate esters	– good fire-resistance
Perfluorohydrocarbons	– the only completely oxygen-resistant lubricants

applications, for example, for the trunnion bearings of steam-heated calenders. Diesters can be used both at lower and high temperatures than mineral oil, but tend to be available only in low viscosity grades. They are used principally for the lubrication of gas turbines, particularly aircraft gas turbines, where advantage can be taken of their low temperature fluidity for starting and their high temperature stability to reduce the need for cooling. Their use in industrial gas turbines reflects the origins of these machines.

The polyalkaleneglycols (PAGs), like mineral oils, are available in a complete range of viscosity grades, both miscible and immiscible with water. The former can be used as fire-resistant fluids; the latter are characterized by having high viscosity indexes and extremely good boundary-lubricating properties. They find particular application in worm gearboxes where there is a high degree of boundary lubrication.

Greases based on liquid synthetic lubricants are also available; their particular application is in the lubrication of high-temperature rolling bearings.

All the synthetic lubricants are more expensive than mineral oil (the perfluorocarbons enormously so), but there are occasions on industrial plant where their use is justified by extending the limits of lubrication. It should be noted that some synthetic lubricants attack the oil-resistant paints and rubber seals used in normal lubrication systems. Special attention has to be paid to this before using a synthetic lubricant.

3.6 SOLID LUBRICANTS

MAJOR SOLID LUBRICANTS

	Temperature limit
ptfe	280°C
Molybdenum disulphide	350°C
Graphite	500°C*

*Low friction depends on adsorbed vapour films; if these desorb, the temperature limit is reduced

No survey of lubricants would be complete without some mention of solid lubricants. The three materials that find some use in industrial applications are graphite, molybdenum disulphide, and ptfe. These materials are

APPLICATIONS OF SOLID LUBRICANTS

- At temperatures outside the range of mineral oils, i.e., at very low temperatures and above 140°C

- In situations where normal fluid lubricants could be squeezed out, e.g.,
  (a) to prevent fretting in joints where the amount of movement is extremely small;
  (b) to prevent seizure in threads that are in heavily-loaded contact for prolonged periods, particularly on boilers and other vessels where they are also exposed to high temperature

- In situations where a fluid lubricant that attracts dirt is undesirable.

essentially boundary lubricants, giving coefficients of friction of about 0.1 under heavily-loaded conditions. When this is compared with the value of 0.01–0.001 that can be realized under hydrodynamic conditions, it will be appreciated that they are no substitute for mineral oils. In fact, it is general experience that within the range of application of mineral oils and greases, additions of solid lubricants give little practical benefit. Their use is to provide lubrication in the circumstances shown on the previous page.

Solid lubricants are available dispersed in oils and solvents, or incorporated in greases. They are best applied as a thin resin-bonded film to a treated surface, e.g., to a phosphated steel surface or an anodized aluminium one.

Table 3.5 (overleaf) gives recommendations for lubricants for different mechanisms on industrial machines.

An Introductory Guide to Industrial Tribology

Table 3.5   Guide to lubricant selection for normal industrial machinery

Component	Condition[1]	Lubricant[2]	Viscosity/ Consistency[3]	Comments
Plain Bearings:				
slow speed (<0.1 m/s) total loss lubricated self-contained	—	Lime or Lithium grease	No. 2	Self-lubricated bearings may be a better choice
self-contained	<60°C	Oil to BS 4475[4]	VG 32-68	Use lower viscosity for disc lubricated,
	>60°C	Oil to BS 489[5]	ditto	higher for ring-oil lubricated
Rolling Bearings (self-contained):				
grease-lubricated	−20 to +70°C	Lithium grease	No. 3	
	100°C max	Lithium complex grease	No. 2	
oil-lubricated	<60°C	Oil to BS 4475[4]	VG 32-68	Oil is required above maximum
	>60°C	Oil to BS 489[5]	ditto	recommended speed for grease
Crankcases	—	Oil to BS 4475[4]	VG 68	
Gearboxes:				
spur and helical	<60°C	Oil to BS 4475[4]	VG 68-220	Choice of viscosity determined by pitch-line velocity
	>60°C	ep oil	ditto	
worm	<60°C	Oil to BS 4475[4]	VG 460	
	>60°C	ep oil	ditto	
Circulation Systems	—	Oil to BS 489[5]	VG 32-68	Use higher viscosity when there are gears in system
Hydraulic Systems	<150 bar	Oil to BS 4475[4]	VG 32	With large oil reservoir (>500 litres)use
	>150 bar	R & O oil with load carrying additive[6]	ditto	R & O inhibited oil[6]

Air Compressors oil-lubricated	<140°C	Oil to BS 4475[4]	Discharge pressure: <10 bar – VG 68 >10 bar – VG220	
	>140°C	Special air compressor oil (DIN 51506)[7]	ditto	
oil-flooded	–	Special rotary air compressor oil	VG 32	
Refrigerant Compressors	–	Oil to BS 2626[8]	VG 32–68	Low pour point oil required when lubricant comes into contact with refrigerant
Internal Combustion Engines	–	Special oil containing dispersant, high temperature, anti-oxidant and possibly VI Improver additives	SAE Viscosity Grade recommended by manufacturer	

Notes
[1] Temperature refers to the bulk lubricant or the discharge temperature in the case of compressors.
[2] Lubricant for more severe condition may be used for easier condition for sake of rationalization.
[3] Viscosity Grade for oils (BS 4231), consistency for greases.
[4] BS 4475: Specification for straight mineral lubricating oils.
[5] BS 489: Specification for steam turbine oils.
[6] R & O oil indicates oil containing rust and oxidation inhibitors.
[7] German standard DIN 51506: Lubricants – Lubricating oils VB and VC without and with additives and lubricating oils VD-L.
[8] BS 2626: Specification for lubricating oils for refrigerant compressors.

CHAPTER 4

Lubricants in Service

Lubricants deteriorate in service by *contamination* and *degradation*.

Small systems (up to 200 litres)   –   change on routine determined by experience.

Large systems (> 200 litres)   –   monitor oil by *visual examination* and routine *laboratory testing*

Major problems caused by contamination with water (emulsification), air (foaming) or solids can be kept under control by proper maintenance.

4.1   CHANGE PERIODS

There are two main reasons for changing lubricating oils: degradation and contamination. The former is caused by oxidation, or additive depletion in the case of additive-containing oils. To obtain a reasonable life from a mineral oil lubricant, the temperature of the bulk oil in the reservoir or sump should not exceed 60°C. On the other hand in systems liable to contamination by water, it is advantageous to maintain the temperature near this limit so that water is evaporated off. The effects of contamination can be reduced or eliminated by filters or centrifuges and by regular draining of separated material from the bottom of the oil reservoir. At some point, however, the maintenance engineer has to decide when a lubricant has reached the end of its useful life. This can be done on the basis of visual examination or by means of laboratory tests.

MONITORING OF OILS IN SERVICE

System size:
< 200 litres   monitoring rarely justified
              change on routine.
> 200 litres   weekly visual examination
              6–12 month laboratory tests

Laboratory testing is expensive and only justified when large quantities of lubricant are at stake, say 200 litres or more. The best practice is to change the lubricant in small systems (< 200 litres) at some convenient period based on experience, as is done with motor engine oils, rather than on tests. It is common experience on industrial plant that oil in well-sealed systems, e.g., the oil in gearboxes operating at normal temperatures, will be perfectly satisfactory for two or three years. For less well sealed systems, such as fabricated oil baths round a chain drive, it may be preferable to change the oil once per year. Table 4.1 gives recommended change periods for small systems.

Table 4.1   Recommended change periods for small systems

System	Temperature	Change period
Poorly sealed systems (e.g., chain drive with fabricated casing)	All	1 year
Well sealed systems (e.g., gearboxes, compressor sumps)	<60°C	2 years
	>60°C	1 year

Experience with grease-packed rolling bearings shows that in all but a few cases on industrial plant the grease should be satisfactory for at least one year, in many cases for 3–5 years. (See Chapter 5 for a method of calculating grease life in rolling bearings.)

For large systems a two-tier monitoring approach is recommended, consisting of a visual examination, at say weekly or monthly intervals, backed up by laboratory analysis at 6–12 months, or as indicated by the visual analysis.

4.2   VISUAL EXAMINATION

The easiest, most useful, and probably the most significant test is a visual examination of the oil; this can be readily carried out on the plant and needs little experience or equipment for its application. A sample of oil from a circulation system, crankcase or gearbox is taken in a clean glass bottle. The sample should, if possible, be taken from the circulating or moving oil; if it can only be taken from a drain point sufficient oil should be run off to ensure that the sample is typical of the bulk oil in the system. The sample should be compared with a sample of the new oil and the previous sample taken for visual analysis. Note that these samples should be kept in a cupboard away from the light; oil is sensitive to sunlight and in time will throw sludge unless kept in the dark. If the oil is clear and does not markedly differ in colour from the retained sample it is passed for further service; if not, it should be stood for an hour, preferably at about 60°C (say, on an office radiator) to allow emulsions or foams to break down and entrained solids to separate out. With dark oils, solids can be detected by carefully inverting the bottle and examining the bottom. The action now to be taken depends on the particular system and may include analytical testing. Where filters or centrifuges are fitted, the presence of water or solids indicates that the cleaning devices are not functioning correctly and attention should be paid to them rather than to the oil. Where there are no purification devices it becomes more difficult to assess whether the oil is still satisfactory. The following points are offered for guidance.

(1) If the oil becomes clear after standing this indicates that when the sample was taken it contained entrained air. (See below for a discussion on air entrainment and foaming and what action should be taken.)

(2) Where a stable emulsion still persists or sludge has settled out, the oil should be changed.

(3) Where water has separated at the bottom of the bottle a check should be made that adequate routine draining from the bottom of the system is being carried out; if persistent water contamination is occurring the source should be looked for and removed. The limit of water contamination for safe operation depends very much on the

circumstances: in systems where the oil lubricates precision parts (e.g., governors, rolling bearings) any free water can cause trouble (water is soluble in oil to about 200 ppm, above this level the oil becomes cloudy); in other systems up to 0.1 percent water is acceptable.

(4) It is not possible to be specific about the proportion of solid contaminants that can be tolerated as this depends very much on the nature of the solids and their particle size. Where there is cause for concern the insolubles have to be determined analytically; an arbitrary figure of 0.2 percent is suggested as a limit. If significant amounts of solids are present it is worth checking with a magnet to see if they are magnetic. The presence of magnetic solids in lubrication systems where there are rolling bearings or gears means almost certainly that wear is taking place and should be investigated. Identifying solids by analysis may also help to indicate the source of contamination. X-ray diffraction analysis is the most helpful in this regard, readily distinguishing between wear products (iron, bronze, white metal), rust (caused by water contamination), or adventitious dirt (silica is normally present).

(5) Severe darkening associated with an acrid smell indicates that the oil is oxidized. Laboratory testing is required to see how far this has gone and whether the oil needs to be changed.

A scheme for visual analysis is given in Table 4.2.

It is important to remember that oil is comparatively cheap, whereas lubrication failures of machines can be expensive; hence if one is uncertain, it may be preferable to change the oil. Visual analysis has the advantage that it gives immediate results and if carried out regularly can alert the need for more sophisticated investigation of the oil condition when large quantities are involved.

Once the principle of visual examination has been accepted and experience gained, there is no reason why it should not also be carried out on small systems after an incident has led to some doubt about the condition of the oil, for example a period of hot running, or possible contamination after hosing down the plant.

Table 4.2 Scheme for visual analysis of used lubricating oils

1 Take sample of circulating oil in clean 2 or 4 oz glass bottle.
2 If oil is dirty or opaque, let it stand for an hour.

Appearance of sample		Reason	Action to be taken	
When taken	After 1 hour		System without filter[1] or centrifuge	System with filter[1] or centrifuge
Clear	Clear	Foaming	None / Cause of foaming to be sought[3]	None / Cause of foaming to be sought[3]
Opaque[2]	Clear oil with separated water layer	Unstable emulsion	Run off water (and sludge) from drain cock[4]	Check centrifuge
	No change	Stable emulsion	Submit sample for analysis	Check centrifuge. If centrifuge fails to clear emulsion, change oil
Dirty	Solids separate[5]	Contamination	Submit sample for analysis	Check filter or centrifuge
Very dark (acrid smell)		Oxidation of oil	Submit sample for analysis	Submit sample for analysis

Notes
1 This includes paper and felt filters, not wire mesh strainers.
2 Both foams (mixtures of oil and air) and emulsions (mixtures of oil and water) render the oil opaque. When an opaque sample is received it should be stood for 1 hour, preferably at 60°C (an office radiator provides a convenient source of heat). A stable emulsion persists after this time; a less stable emulsion shows a separated layer of water below the oil; a foam breaks down, liberating the gas and leaving a clear oil.
3 Foaming is usually mechanical in origin, being caused by excessive churning, the use of unnecessarily high pressures in spray systems for lubricating gears, the impingement of pressurized return oil on the surface of the reservoir or air leaks in the oil pump suction; the necessary action should be taken to prevent it. Foaming can also result from minor quantities of certain contaminants, e.g., solvents remaining from degreasing, grease, etc.; in this case the oil should be changed.
4 Failure of water to separate from the oil in service may be the result of inadequate lubricant capacity or of the oil pump suction being too close to the lowest part of the reservoir; this can only be cured by the appropriate modification. More commonly, it results from the re-entrainment of separated water from the bottom of the sump or oil reservoir, when, by neglect, it has been allowed to build up in the system.
5 In a dark oil, solids can be seen by inverting the bottle and looking at the bottom.

4.3 ROUTINE LABORATORY TESTING

In addition to visual examination and any analytical testing that arises from it, routine analytical tests should be made on the oil in large systems to detect more subtle changes than can be detected visually. In general, these changes are slow and routine analytical testing need be no more frequent than every six or twelve months. This service is usually carried out by the oil supplier or a specialist contractor. The maintenance engineer should, however, know something of the significance of the tests in order that he can assess the results.

ROUTINE LABORATORY TESTS

	Limit
Viscosity at 40°C	±15 percent of new oil value
Acid value:	
straight oils	2 mg KOH/g
turbine-type oils	1 mg KOH/g
aw hydraulic oils ⎫ ep gear oils ⎬	Discuss with supplier
Water	200 ppm

4.3.1 Acid value (neutralization number)
The main change that takes place in the oil is gradual oxidation. This shows itself by darkening of the oil, the development of acidity, an increase in viscosity and, eventually, if allowed to proceed, deposition of sludge and solidification. The oxidation reaction is autocatalytic, that is, once it starts the oxidation products themselves catalyse the oxidation reaction so that it proceeds at an increasing rate (Fig. 4.1). The acids formed are not corrosive to normal construction materials, but measurement of the acidity gives a useful guide to the condition of the oil. A new oil should have an acid value <0.05 mg KOH/g. Straight mineral oils should be changed when the acidity reaches a value of 2–3 mg KOH/g: the

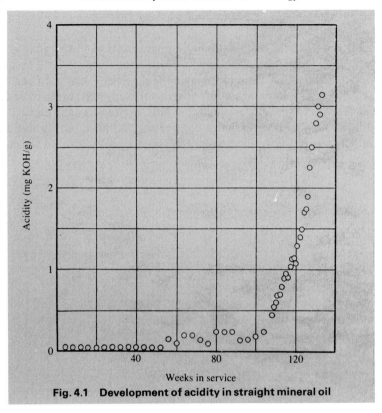

Fig. 4.1   Development of acidity in straight mineral oil

exact value is not critical, but experience shows that the useful life of an oil is limited once this value is reached.

In oils containing oxidation inhibitors, there is a long induction period before acidity develops, and it is found that the rate of increase gives a much better indication of oil deterioration than the actual value of acidity. Thus with inhibited oils the value should be logged and the oil changed when a definite increase is detected; a limit of 1 mg KOH/g should be used with these oils. It should be noted that certain gear and hydraulic oils may have initial acid values 0.5–1.5 mg KOH/g; this is caused by the load-carrying additives. In such oils the acid value will initially fall and then begin to rise; these oils should be changed when a rise in the acid value is detected, particularly when it rises to 1 mg KOH/g.

4.3.2 Viscosity

As shown in Chapter 2 most mechanisms are relatively insensitive to changes in the viscosity of the lubricant. However, it is useful to include a determination of viscosity in the routine analysis. Primarily, a change in viscosity may indicate that an incorrect grade has been used in topping up or that there has been significant contamination by oil-soluble materials. Secondly, an increase in viscosity coupled with the development of acidity indicates oxidation. A change in the viscosity at 40°C of more than 15 percent compared with the new oil requires further investigation.

4.3.3 Additive depletion

Depletion of additives in service is not normally a problem with an annual top-up rate of 5–10 percent. If a check of the additive level is required because of exceptional circumstances (e.g., chemical contamination), suitable tests and limiting values should be discussed with the oil supplier.

It cannot be emphasized too strongly that the analytical results are only representative of the sample. It is most important, therefore, that this should be taken carefully and that a clean container should be used if meaningful results are to be obtained and expensive analytical costs are not to be wasted. The sample should be at least 250 ml; the bottle should be sealed and adequately labelled.

4.3.4 Testing of grease

In general, it is not necessary to carry out routine testing of grease, though on certain occasions visual examination may be useful in assessing the condition. In normal circumstances grease darkens in service and may change in consistency, becoming either harder or softer. Grease may also become softer, and could even liquefy if mixed with a grease of a different type. If overheated, it oxidizes like oil and hardens, eventually becoming rubbery and of no value as a lubricant. Where fretting damage occurs with ferrous surfaces the grease takes on a characteristic reddish colour.

4.4   MAINTENANCE OF LUBRICANTS IN SERVICE

Apart from contamination, few problems arise with mineral oil lubricants in service. In particular circumstances the following aspects may require attention (see Chapter 7 for more details).

Contamination by water (emulsification)	– free water (>200 ppm total water) causes corrosion and creates conditions for biochemical growth
Contamination by solids	– level of solids has to be controlled by filtration
Contamination by gas (aeration and foaming)	– instances are not common, but when they occur are a difficult problem

4.4.1   Contamination by water

Oils in good condition contain up to about 200 ppm of water in solution. Above this level the oil becomes cloudy, though generally it separates readily if the oil is allowed a period of quiet, for example a residence time of 5–10 minutes in the reservoir. However, if oils degrade or become contaminated, the rate of separation is reduced and even stable emulsions may be formed. It is undesirable to operate with oils in this condition as corrosion may occur and the corrosion products (iron hydroxides) act as emulsion stabilizers. If centrifuges or coalescers are fitted, then these should be checked for satisfactory operation. The usual faults are (a) too low temperature in centrifuges (the oil should be heated to 80°C for satisfactory removal) or (b) blinding in the case of coalescers through failure to drain. Even without purification equipment, failure to drain settled water from the oil tank may result in emulsion formation when the water is taken up by the pump suction. When stable emulsions are formed the oil should be changed as soon as possible.

A second effect of water contamination is that it can create the conditions for the growth of micro-organisms. These organisms (yeasts, other fungi, and bacteria) are nourished by the lubricant hydrocarbons or lubricant additives, but depend on the presence of free water. The problem with biological contamination is that it can result in the stabilization of emulsions, leading to the blocking of fine clearances, and in the creation of corrosive conditions in the lubrication system.

4.4.2 Foaming

Most lubricated mechanisms tend to generate bubbles of air in the oil, either through the intimate mixing of air and oil or by the release of dissolved air when pressures are let down. Lubrication systems should be designed to minimize the generation of bubbles and allow for their complete separation. Foaming problems, which give rise to unsatisfactory lubrication conditions, arise when this does not occur.

It is important to realize that two quite separate physical phenomena are involved, i.e., detrainment of dispersed bubbles and bubble collapse at the surface, and that the factors governing them are completely different. Detrainment is affected by oil viscosity (a good reason for maintaining the oil in the reservoir at about 60°C) and quality, while bubble collapse depends on the surface-active properties of the oil.

Entrainment and foaming problems arise in lubrication systems either because excessive air is entrained or the rate of foam collapse is too low. Mechanical causes of air entrainment can arise in a number of ways: air leaks into the oil pump suction; rough edges to delivery nozzles; splashing return oil on to the reservoir surface; or whipping of loose pieces in the surface of the oil. Slowing down of bubble detrainment and foam collapse is the result of degradation or contamination of the oil; it is rarely possible to determine the cause by analysis as only very small amounts (ppm) of the contaminant need to be present. (See Chapter 7 for details of the lubricant system design that affect air entrainment and reduce the risk of foaming.)

In small systems the best approach is to change the oil; if this does not improve the situation then it is necessary to look for a mechanical cause. When large quantities of oil are at stake, comparative tests on a sample of the new oil and oil from the system can be carried out to check which of the three possible mechanisms are responsible. Table 4.3 provides a suitable scheme.

4.4.3 Contamination by solids

Oil circulation systems should be fitted with filters to control the level of solid contaminants. For systems containing bearings and gears a filter rating of 15–25 μm gives adequate protection. For systems in which there are fine clearances (e.g., hydraulic systems, turbine governor control systems) the filter rating should be 5 μm or less.

Table 4.3   Identification of causes of aeration and foaming

Oil sample analysis		Cause of poor air release or excessive foaming
Air release[1]	Foam stability[2]	
No change from new oil	No change from new oil	Mechanical fault
Significant increase from new oil	No change from new oil	Contamination by silicone[3]
Significant increase from new oil	Significant increase from new oil	Contamination by basic material[4]

Notes
[1] Test method IP313.
[2] Test method IP146/ASTM D892.
[3] As little as 2 ppm of a silicone can slow down the rate of air release. Plastic sealants and gaskets are possible sources of silicones.
[4] Possible sources are diesel engine oils and pipe lagging.

CHAPTER 5

Selection of Bearing Type

Choice of bearing primarily determined by operating speed and load:

Self-lubricating and
  rubbing bearings – low speeds, low loads
Rolling bearings – moderate speeds, moderate loads
Fluid-film bearings – high speeds, high loads
White metal bearing alloys used in most industrial fluid film bearings with theoretically infinite life:

Failure mechanisms: wiping
fatigue
scoring

A bearing is a mechanism where load is transmitted between relatively moving parts. There are five basic designs in which this can be done.

(1) Rubbing bearings in which one surface slides over the other
(2) Rolling bearings in which one surface rolls over the other
(3) Fluid film bearings in which the surfaces are separated by a thin film of fluid (externally pressurized or hydrodynamically generated)
(4) Magnetic bearings in which the surfaces are kept apart by magnetic forces
(5) Flexure pivots in which parts are connected by a spring or an elastic member.

It is not proposed to deal with hydrostatic bearings or bearings separated by repulsive magnetic forces as these are all somewhat specialized and do not yet find much application in industrial machinery, although the former are being increasingly used in machine tools. The great advantage of the hydrostatic bearing is its ability to control the stiffness of the lubricant film and its low value of coefficient of friction over the complete range of operating conditions, including starting. It should, however, be noted that this is only an advantage in reducing the losses at the point of application of the load. From the point of view of overall efficiency the power consumption in supplying the lubricant to the bearing has to be taken into account.

Magnetic bearings are complicated in design and are only used in rather special applications in which there is a strong incentive to eliminate the use of oil.

Flexure pivots are limited to oscillating movements and are particularly useful as no lubricant is required since no relative sliding takes place. They should be considered for linkages in place of lubricated pin joints, particularly in high temperature or chemical environments where lubrication is a problem. There are basically two types: those in which movement takes place elastically in a rubber member and those in which crossed springs bend to allow the movement. The limit of use is set by the fatigue limit of the elastic member. Figure 5.1 shows the loads that can be transmitted in terms of the angle of oscillation. For a fixed angle of oscillation rubber bushes can carry significantly higher loads; crossed spring pivots can be designed with a much lower angular stiffness for a given load capacity and are much more tolerant of a wider range of environmental conditions (see Engineering Sciences Data Unit, Data Item 67021).

The vast majority of bearings for industrial machines are chosen from one of the first three types in the above list. Each has its strengths and weaknesses and some consideration has to be given to these in order to select the most appropriate type for a particular application.

The basic parameters to be considered in selecting the bearing type are speed and the load to be transmitted. These can be considered as an initial guide, although other environmental factors such as temperature, contamination, and so on may be decisive in determining the actual type from those best suited to meet the load and speed conditions.

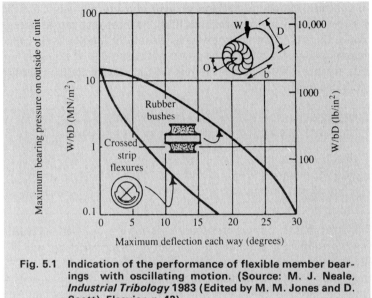

Fig. 5.1 Indication of the performance of flexible member bearings with oscillating motion. (Source: M. J. Neale, *Industrial Tribology* 1983 (Edited by M. M. Jones and D. Scott), Elsevier, p. 42)

5.1 SELECTION OF THE BASIC BEARING TYPE

Consideration of the mechanism of operation of the different types of bearings shows that they have different load–speed characteristics.

Dry rubbing bearings are limited either by wear or overheating, leading to seizure or loss of mechanical properties. To a first approximation life is inversely proportional to the product of specific load p (load/unit area) and surface velocity v ($L \propto 1/pv$) with limiting values of load and speed determined by mechanical strength and thermal considerations, respectively.

The limiting factor in the life of rolling bearings is fatigue of the rolling elements or raceways. As will be shown in Chapter 6 the fatigue life of rolling bearings is inversely proportional to the product of load (W) to the power x (where $x = 3$ for ball bearings, 10/3 for roller bearings) and speed (n): $L \propto 1/W^x n$. Again there is a limit on the maximum load that can be carried by the bearing without permanent plastic deformation at the contact points.

In contrast, the operating limit of hydrodynamically-lubricated bearings is determined by the minimum film thickness that can be accepted. Provided there is an ample supply of lubricant, the load-carrying capacity increases as some function of the power of the speed, decreasing at higher speeds because it is not possible to remove the heat; hence the viscosity of the lubricant begins to fall.

RELATIONSHIP BETWEEN LOAD-CARRYING CAPACITY AND SPEED IN JOURNAL BEARINGS

Rubbing bearings	$W \propto n^{-1}$
Rolling bearings	$W \propto n^{-1/x}$
Hydrodynamic bearings	$W \propto n$

Oil-impregnated, porous-metal bearings have properties intermediate between a rubbing bearing and an oil-film lubricated bearing, due to the limited supply of oil.

In industrial engineering applications loads extend over seven orders of magnitude; speeds over six. It is thus convenient to use logarithmic scales to compare the load–speed characteristics of the different types of bearings, particularly if one is looking for broad guidance rather than detailed accuracy. Guidance charts, based on the above ideas, have been prepared by M. J. Neale (see Engineering Sciences Data Unit, Data Items 65007 and 67033). The load–speed guidance for journal bearings is reproduced in Fig. 5.2. This is based on a life of 10,000 h, taking reasonable values for the wear rates of rubbing bearings and the manufacturers' data for oil-impregnated and rolling bearings. Hydrodynamically-lubricated bearings have been included by assuming a reasonable practical minimum oil-film thickness. Rotational speed is used for the ordinate so that for any particular shaft diameter it is possible to select the bearing type that will give the maximum load-carrying capacity (indicated by the heavy line in the diagram) for a particular rotational speed. Ultimate limits are set for the speeds at which rolling bearings can be used by centrifugal effects and to all bearings by the burst limit for a steel shaft.

The load–speed guidance charts have two main functions for the designer: first, to give initial guidance on the type of bearing most likely to be suitable with a new design; and secondly to check that when an existing

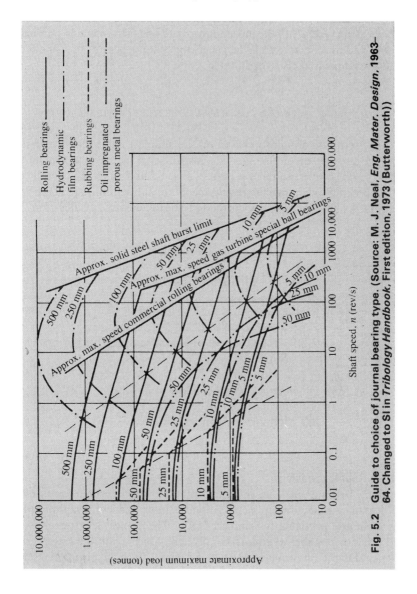

Fig. 5.2 Guide to choice of journal bearing type. (Source: M. J. Neal. *Eng. Mater. Design*. 1963–64. Changed to SI in *Tribology Handbook*. First edition. 1973 (Butterworth))

design is being extended (increased in size, speed, or rating), the presently-used bearing type is still the most appropriate one. The machine purchaser should find the charts useful when checking that the bearing design offered with a machine is likely to be suitable.

5.2 MORE DETAILED CONSIDERATIONS IN BEARING SELECTION

Clearly, selection should be based on the simplest and cheapest bearing that will provide the design life. The bearing cost increases as one moves from the bottom left to the top right of Fig. 5.2 and when the overall cost of the bearing, including the requirements for lubrication, are taken into account, six bearing arrangements have to be taken into consideration.

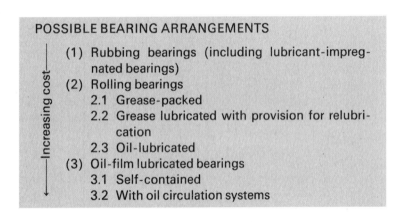

POSSIBLE BEARING ARRANGEMENTS

Increasing cost →

(1) Rubbing bearings (including lubricant-impreg-
nated bearings)
(2) Rolling bearings
 2.1 Grease-packed
 2.2 Grease lubricated with provision for relubrication
 2.3 Oil-lubricated
(3) Oil-film lubricated bearings
 3.1 Self-contained
 3.2 With oil circulation systems

5.2.1 Rubbing bearings

Rubbing bearings may be of metals that are run dry or intermittently lubricated, mixed powder–metal/graphite sinters, carbon graphite, plastics including particularly reinforced ptfe. The choice is extremely wide and it is not possible to go into details here. Once again the Engineering Sciences Data Unit provides an extensive guide, not only on material selection but also on design, and this should be consulted for further information (Engineering Sciences Data Unit, Data Item 87007). Figure 5.3 gives

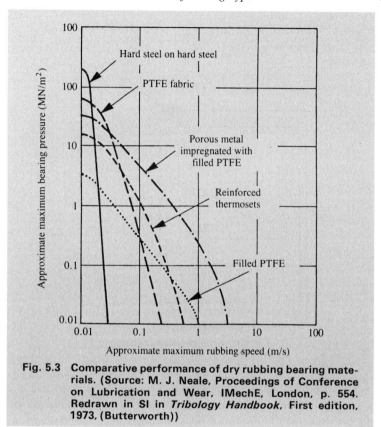

Fig. 5.3 Comparative performance of dry rubbing bearing materials. (Source: M. J. Neale, Proceedings of Conference on Lubrication and Wear, IMechE, London, p. 554. Redrawn in SI in *Tribology Handbook*, First edition, 1973, (Butterworth))

a load–speed diagram for a range of typical materials; it will be seen that few of these conform to the simple inverse load–speed relationship.

A particular problem with plastics materials is their high coefficients of thermal expansion and generally poorer dimensional stability compared with metals. This means that it is necessary to use high clearances. A simple way in which this can be overcome is to use a thin layer of the plastic on a steel backing. Examples of this are the Glacier DU Bearing that consists of a sintered bronze layer impregnated with ptfe and lead on a steel backing, or the Glacier DX Bearing that uses a grease-impregnated

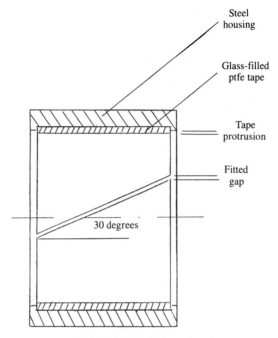

Fig. 5.4   Split ptfe tape bearing

polymer supported on a steel backing. An alternative is to use a split reinforced ptfe tape (1–1.5 mm thick) confined in a steel housing, giving negligible expansion in the radial direction while accommodating the circumferential expansion at the split (Fig. 5.4).

5.2.2   Rolling bearings
The selection of rolling bearings is considered in more detail in Chapter 6. The majority of rolling bearings are manufactured from hardened and tempered low chromium steel, with case-hardened steel for the larger sizes. Tempering temperatures are about 140°C and normal bearings will suffer a loss of load-carrying capacity if used above the tempering temperature. Bearings with special steels are available and should be used for high temperature applications. The transition from grease-packed bearings to grease-lubricated bearings with provision for relubrication and oil-

lubricated bearings is reflected by an increase in cost. Grease life is the determining factor in choice: a method for calculating grease life is given in Chapter 6. Recommended limits for the two arrangements of grease lubrication are given in Table 6.4. Limits for oil lubrication are given in the bearing manufacturers' catalogues.

5.2.3 Hydrodynamically-lubricated plain bearings

The upper limit to the use of self-contained bearings is determined by heat removal and is largely a matter of speed. Recommended limits are given in Chapter 7.

Bearings for industrial machines are commonly made in split halves, with either an axial oil groove on the split line, extending over 75–80 per-cent of the axial width, or with a central circumferential groove in the case of bearings subject to rotating loads (e.g., main bearings for reciprocating compressors; in both cases the oil is fed from the back of the bearing shell. While the industrial user is not normally concerned with the detailed design, two points are worth mentioning. In the case of bearings with axial oil-feed grooves, it is important to cut dirt gutters at the ends of the grooves to ensure that any dirt circulated with the oil is flushed out of the bearing (Fig. 5.5). In the majority of machines the bearings are installed with the split line horizontal. In the case of gears, however, the bearing should be rotated so that the resultant load line is at the centre of a bearing half and does not fall near the split line. If the machine is subsequently uprated it may be necessary to compensate for this by adjusting the angle of rotation of the bearing shells.

Industrial users are not normally concerned with the design of plain bearings; there are, however, a number of factors, mainly concerned with material, of which the user should be aware.

The choice of materials for oil-film lubricated bearings is determined mainly by the requirement to avoid damage during start-up or momentary periods of interruption of the oil-film. Also, a precise balance is required between the strength to support the load and sufficient deformability to ensure the conformability with the opposing surface and the ability to accommodate minor misalignment. Another property of considerable importance is embedability, i.e., the ability to absorb hard contaminant particles within the body of the material so that they do not score the journal or thrust collar.

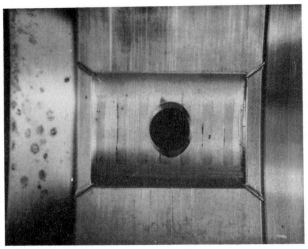

Fig. 5.5   Oil groove with dirt gutters

Bearing metals normally consist of two major metallurgical constituents: a soft phase to give conformability and embedability and a hard phase to give the necessary strength to carry the load. The alloys outlined below are in widespread use.

BEARING MATERIALS

- White metals:
  basically alloys of tin and lead, usually with antimony and copper
- Bronzes, particularly lead-bronzes
- Copper–lead alloys
- Aluminium–tin alloys

White metals (Babbitts) have all the desirable properties of bearing metals and are by far the most important bearing materials in industrial machines, being limited only by their strength (fatigue resistance to alternating loads) and reduction of yield strength above 120°C. White metals have low melting points, about 240°C: this means that in the event of failure it may be possible to shut down the machine without damage to the rotor, the molten metal acting as an emergency lubricant. This is a major factor in the philosophy of bearing design for large and important machines; the bearing is a self-contained component that can be replaced relatively easily in the event of failure. If damage to the more expensive rotating parts can be avoided, repair of the machine after a failure, with only the bearing to be replaced, should thus be a comparatively simple and inexpensive operation. The need for simple and regular bearing replacement on high-speed rotary machines, either because of wear or fatigue arising from out-of-balance, has to be accepted. The remaining bearing alloys are either higher-melting or harder and are much more likely to damage the rotating parts in the event of failure.

White metals have insufficient strength and are normally used as a lining on a steel or bronze shell. The fatigue strength can be increased by using very thin linings on a steel shell. Thin-wall bearings with a white metal thickness of 0.10–0.15 mm are widely used on engine bearings. However, in industrial machines there is some merit in using white metal layers of 0.5–3.0 mm thickness, to provide good conformity and embedability as well as sufficient white metal to cover an emergency shut-down after failure, the maximum thickness being determined by the minimum clearance of the rotating parts in the casing so that rubs are avoided. Thick wall bearings have been used to allow the final alignment of the machine by scraping in. This practice is, however, no longer acceptable, since the advantages of using pre-machined bearings that can be fitted directly outweigh any additional manufacturing cost.

A very wide range of white metals is available, going from the high-tin alloys (ca. 90 percent tin) to the high-lead alloys (ca. 80 percent lead). As far as thick-wall bearings for industrial machines are concerned the requirements are fully met by Grades C and G from the British Standard BS 3332. The composition and properties of these alloys are summarized in Table 5.1.

It will be seen that the tin-rich and lead-rich alloys have virtually identical properties. In general, Grade C is preferred because it is slightly easier to bond to steel than Grade G. Grade G is more resistant to attack by active sulphur compounds and is preferred in cases where this can be a problem (see Chapter 9).

Table 5.1   White metals for industrial thick-wall bearings

BS 3332 Grade*	Composition Main alloying elements (%)					Solidus °C	Hardness HB		
	Sn	Sb	Cu	Pb	Cd		25°C	50°C	100°C
C	86	9	4	0.35	0.05	239	30	23	14
G	5.5	15.5	0.7	78	0.05	242	26	21	14

* As far as application is concerned Grade C is equivalent to the German DIN. Grades WM90 (90% Sn) and WM80 (80% Sn), Grade G to WM5 (75% Pb).

A low carbon mild steel is the preferred material for the bearing shell. Because of its porosity cast iron leads to problems in remetalling as the oil that soaks into the cast iron is difficult to remove and this can make it difficult to obtain satisfactory bonding. Low alloy steels are not required and should not be used as it is more difficult to obtain good bonding than with plain carbon steels.

The normal practice is to coat the steel backing with a layer of pure tin before casting on the white metal lining. Very satisfactory metallurgical bonds can be achieved between steel and white metal in this way. The use of dovetails is not only unnecessary, but can be positively harmful in that they can set up uneven thermal gradients during cooling that can give rise to stresses that disrupt the metallurgical bond. Even coarse gramophone finishes are unnecessary and make it more difficult to check that good bonding has been achieved. Ultrasonic testing can be used to check bond integrity (see ISO 4386–1). The wide range of standards in use suggests that the limits are not very critical. Using a 10 mm twin-crystal probe, a minimum of 85–90 percent bonding, with the maximum permissible length of any individual flaw not greater than 5–10 mm, should be satisfactory.

WHITE METAL BEARINGS

Backing metal	Low carbon steel
Bonding	Pure tin bond layer
	(dovetails should not be used)
Integrity	Check by ultrasonic testing

No direct relationship has been established between the mechanical properties of white metals and their load-carrying capacity. To some extent this is because the true loading is not known and, moreover, differs at starting when the surfaces are in solid contact, and under running conditions, when they are separated by a liquid lubricating film and the load is spread over a large area.

By convention, the loading is defined by the load per unit projected area and the performance of different bearing metals and designs is established on the basis of simulative laboratory testing.

The start-up loading for white metal is limited to about 3.5 N/mm^2. Higher loadings can be accommodated by the use of hydraulic jacking, but this is an added complication that is only justified in the case of very heavy rotors. Under running conditions with a hydrodynamic lubricating film the loading can be increased to about 10–15 N/mm^2, the limiting value depending on the bearing design and the nature of the load, steady or alternating. At higher loadings than can be carried by white metals, the choice lies with copper or aluminium–tin bearing alloys.

Both lead–bronze and sintered copper–lead bearings are used. The use of the latter is confined to thin-shell bearings for automotive applications. The lead–bronzes are widely used as solid bushes in heavily-loaded industrial machines with lead contents ranging from 5–36 percent; the higher the lead content the lower the strength, but the better the conformability and embedability. Three typical alloys covered by BS 1400 are shown in Table 5.2. Under marginal conditions the higher lead compositions should be used; for example in the little-end bearings of single-acting reciprocating compressors where, because of the small oscillating movement and absence of load reversal, lubrication is almost entirely in the boundary regime. At high loadings, however, there can be a tendency for extrusion of the lead if the highest lead composition is used and the

Table 5.2 Lead–bronze bearing alloys

BS 1400 Grade	Composition (%)			Solidus °C	Hardness HB 20°C
LB4C	85	5	10	ca. 325	45–78
LB2C	80	10	10	ca. 325	65–90
LB5C	75	5	20	ca. 325	45–70

80:10:10 composition is the preferred bearing material for most applications. Successful operation can be obtained with peak loadings of 15–20 N/mm². It is necessary to use harder journals than with white metals; 250 HB minimum is generally required with the lead–bronzes.

A further improvement in strength is obtained with the aluminium–tin alloys. The maximum satisfactory structure obtainable by normal metallurgical manufacturing methods is about 7 percent tin; with higher tin contents the mechanical properties fall off as the tin tends to become concentrated at the grain boundaries. This problem has been overcome by mechanical working and both steel-backed and solid aluminium–tin bearings with 20 percent and 40 percent tin (reticular aluminium–tin alloys) are available. The reticular aluminium–tin alloys, which have good high temperature properties, are particularly resistant to fatigue and cavitation. With their high tin content they have reasonable embedability and do not require the fine degree of filtration that is necessary with the 7 percent tin alloys, nor do they require the use of hardened journals.

5.3 PLAIN BEARING FAILURES

WHITE METAL BEARING FAILURES

Wiping	–	breakdown of lubricant film
Fatigue	–	excessive vibration or reciprocating loads
Premature fatigue	–	poor bonding
Wire-wool failure	–	occurs with high-chrome journals or thrust collars

If insufficient oil is fed to the bearing to remove the heat generated, the temperature rises, thereby reducing the viscosity of the oil and ultimately melting the lower melting constituent of the bearing metal. This gives rise to a form of failure known as wiping. It should be recognized that, in the case of high-speed bearings, high enough temperatures can occur in the oil film to cause wiping without a breakdown in the oil film having happened. This is particularly the case with the low melting white metals and, as already mentioned, can be useful in preventing damage to journals in a crash shut down. With lead bronzes this type of

Fig. 5.6   Wiping of white metal bearing

failure may lead to wiping of the lead, and extrusion of lead from the back of the bearing, but with the higher temperature it is seldom possible to avoid damage to the journal. This is a major reason for using white metal bearings wherever possible. Figure 5.6 shows wiping in a white metal journal bearing.

A second common failure type is fatigue caused by excessive alternating loads arising from out-of-balance or bearing instabilities. This causes cracking of the white metal, leading to the creation of loose pieces that are held in position because of the low clearance space (Fig. 5.7) or are eventually broken up and carried away in the oil. Poor bonding of the white metal reduces its fatigue strength and gives rise to premature fatigue failure. Figure 5.8 shows an example of premature fatigue where the use of dovetails had led to poor bonding. The two cases are readily distinguished. Where fatigue has been caused by poor bonding the cracks run normally to the surface and penetrate to the steel backing. With well bonded white

Fig. 5.7   Fatigue failure of white metal bearing

metal the crack runs entirely within the white metal and the piece breaks
away leaving the tinned layer on the steel intact (Fig. 5.9).

Another type of failure, known as wire-wool failure, is associated with
the use of chromium steel journal and thrust collars and is not confined to
white metal bearings. This is a dramatic failure, grooves up to 5 mm or
more in depth being machined into the journal or thrust collar (Fig. 5.10)
in a matter of seconds and the bearing housing filled with a mass of wire
wool, surprisingly often with comparatively little damage to the bearing
(Fig. 5.11). In industry these failures have been met with steels containing
from 3–18 percent chromium operating at surface speeds above 15 m/sec.
The failure is initiated by a hard particle embedding into the soft bearing
metal and bridging the oil gap. At high rubbing speeds material from the
journal is transferred and, at the high temperature generated, this is
carburized by the oil producing hard chromium carbide, which then acts
as a cutting tool on the opposing bearing surface.

Fig. 5.8 Premature fatigue of white metal. Poor bonding caused by oil
absorbed in cast iron shell

5.4 PROCESS FLUID LUBRICATED BEARINGS

The use of the process liquid for lubricating the bearings can be advantageous in glandless pumps by avoiding sealing problems with toxic or
hazardous fluids (e.g., high-temperature heat transfer fluids), or with
multistage pumps by increasing the shaft stiffness through reduction of the
bearing separation.

BEARING MATERIALS FOR PROCESS FLUID LUBRICATION

Carbon-graphite	Organic liquids
Steel-backed ptfe	Organic liquids
Reinforced ptfe strip	Organic liquids
Reinforced pf resin	Clean water
Hard rubber	Aqueous solutions, silty water
Silicon carbide	Aqueous solutions, acids

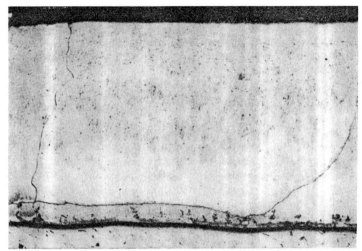

Fig. 5.9   Fatigue crack running down through the white metal and
then along, but above, the bond layer that appears as a
broad black line. (Source: D. F. Wilcock and E. R. Booser,
Bearing Design and Lubrication, 1957 (McGraw Hill, New
York))

The problem with most process fluids is that they are deficient in
boundary lubricating properties. This can be compensated for by using
bearings with self-lubricating properties. Carbon–graphite has been used
satisfactorily with heat transfer fluids, steel-backed ptfe, or ptfe strip with
light hydrocarbons, and phenol formaldehyde resins with aqueous liquids.
In practice it is found that start-up loads should be limited to about 0.25–
0.5 N/mm² with these materials.

Hard rubber cutless bearings with helical grooves are very effective for
water applications, particularly water containing silt that rolls round the
bearing and is then washed away along one of the grooves. They are used,
for example, in submersible water pumps.

Silicon carbide running against itself has been used satisfactorily with
corrosive materials such as concentrated nitric acid and with water in
multi-stage boiler feed pumps.

Fig. 5.10   Wire wool fatigue showing damage to the shaft

Fig. 5.11   Wire wool failure showing damage to the bearing

CHAPTER 6

Rolling Bearings

- Rolling bearings are available in extensive ranges of types and sizes with single-row radial ball and cylindrical roller bearings preferred.
- Design is based on fatigue life.
- Grease provides the simplest lubrication arrangement, limited only by grease life.
- Identification of failure mechanisms:
  fatigue
  premature fatigue
  cage wear
  surface distress
  plastic deformation
  useful in fault diagnosis.

The use of sliding, either with the surfaces in solid contact (dry or boundary lubricated) or separated by a fluid film is one way of transmitting load across relatively moving surfaces; an alternative is to roll one surface over the other.

6.1 DESIGNING WITH ROLLING BEARINGS

Rolling bearings offer a number of real advantages to the designer. They are available as complete items ready to be fitted to the shaft and housing with full details available on load-carrying capacity, operating limits, fitting requirements, lubrication arrangements, sealing, and so on. This eliminates the detailed design that is necessary with plain, hydro-dynamically-lubricated bearings. Further, both radial and axial loads can be carried by the one bearing. Finally, the lubrication arrangements can be very simple: the grease-packed rolling bearing, that can be left for long periods without attention (two years or more) and does not require elaborate

95

provision for sealing, can be used for a very wide range of applications. Rolling bearings are not critical of lubricant viscosity, so they can readily be installed inside machines that are filled with lubricant, for example gearboxes, without the need for a separate lubrication system.

These features of rolling bearings mean that they can frequently provide the cheapest bearing design. However, they do have certain limitations and due attention must be paid to these to ensure that, because of the cheapness and reduced design effort associated with them, they are not used beyond the limits of their reliability in industrial machines.

Unlike plain bearings, which have no limit to their life if properly designed and used, rolling bearings have a definite life. This arises because of the repeated stresses that occur between the rolling elements and raceways at the loaded points of contact, the surface of which breaks up by a fatigue mechanism with repeated stressing. Life is a function of load and the number of revolutions. Being a fatigue process the life of a bearing under specified conditions of load and speed can only be predicted statistically. Testing large numbers of bearings at a prescribed load and speed gives the life relationship illustrated in Fig. 6.1.

For design purposes in industrial machines the 'life' is taken as the period when ten percent of the bearings have failed. This is known as the L_{10} life and is normally expressed in 10^6 revolutions. For machines running at constant speed the life in hours, the $L_{10(h)}$ life, is generally used. The 'average life', namely the period when fifty percent of the bearings would have failed, is about five times the L_{10} life, and individual bearings may survive many times longer than this. Bearing life distribution curves were originally determined experimentally. Empirical methods for calculating rating life, based on bearing geometry, are now available (see, for example BS 5512/ISO 281: specification for rolling bearings – dynamic load ratings and rating life. Part 1 – Calculation methods).

ROLLING BEARING LIFE

L_{10} *life.* The number of revolutions that 90 percent of an apparently identical group of rolling bearings operating under the same conditions exceed without evidence of fatigue.

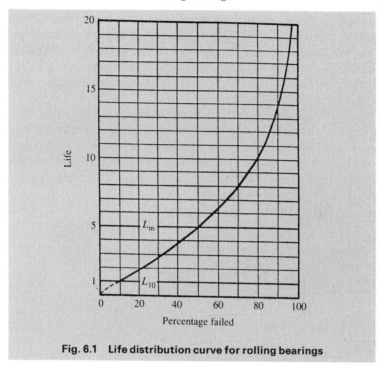

Fig. 6.1 Life distribution curve for rolling bearings

Rolling bearings operate with coefficients of friction similar to those that can be obtained under normal hydrodynamic lubrication conditions; this is because the majority of rolling bearings operate with an elasto-hydrodynamic lubrication film at the rolling contacts. This was not appreciated at the time that the datum for fatigue for rolling bearings was being generated experimentally and, in order to obtain the results in reasonable test periods, testing was carried out at high loading conditions. At such conditions breakdown of the elasto-hydrodynamic lubrication film occurred and the rolling surfaces were in solid contact. Analysis of rolling contact with elasto-hydrodynamic lubricating films shows that the maximum pressure in the film is below the fatigue strength of rolling bearing steels and hence fatigue should not occur. It would thus seem inappropriate to apply L_{10} life data generated from solid contact conditions to normally operating bearings, though it is a matter of experience that such bearings eventually fail by fatigue. The reason that failure occurs is that most lubricants contain solid particles of the same order of size as the thickness of the elasto-hydrodynamic film. On passing through the contact these particles bridge the film and create local stresses high enough to cause eventual fatigue failure.

Years of operating experience have established $L_{10(h)}$ lives that are appropriate for use in the design of rolling bearings for different applications and, while the derivation of $L_{10(h)}$ life is based on somewhat dubious grounds, such $L_{10(h)}$ lives do provide a useful practical criterion for design. Table 6.1 (taken from Engineering Sciences Data Unit, Data item 81005) gives typical values for use in design.

L_{10} life is given by the following relationship.

L_{10} LIFE

$$L_{10} = \left(\frac{C}{P}\right)^{x} \text{ revolutions}$$

for constant speed:

$$L_{10(h)} = \frac{L_{10}}{60n}\left(\frac{C}{P}\right)^{x} \text{ hours}$$

P = Equivalent Radial Load
C = Basic Dynamic Capacity of the bearing,
x = 3 for ball bearings, 10/3 for roller bearings.
n = rotational speed in rev/min

Values for C are given in the bearing manufacturers' catalogues, together with the method of calculating the equivalent radial load for a combination of radial and axial load for different bearing types.

There have been considerable improvements in the manufacture of rolling bearing steels since the original fatigue life data were obtained, giving enhanced resistance to fatigue. This, together with the understanding of the lubrication conditions at the rolling contacts, has led to the incorpora-

Table 6.1   Specified life of rolling bearings for common applications

The following is a sample of the usual recommended values of the basic life for different types of machines and applications. Further values can be found in the bearing manufacturers' catalogues. However, the designer should be aware of the fact that the recommended life values given in this table originated before life adjustment factors came into general use and allowance has already been made for contingencies such as high reliability requirements. The designer should, therefore, exercise caution in applying the recommended life values to avoid overspecifying the bearing size. Experience has shown that these figures are realistic without the use of life adjustment factors.

Class of machine	$L_{10(h)}$*
Machines infrequently used: hand tools and household appliances.	500

Machines used for short periods or intermittently and
    whose breakdown would not have serious
    consequences:
    lifting tackle in workshops, foundry cranes, jib cranes.
	500–2 000

Machines working intermittently whose breakdown
    would have serious consequences:
    electric motors for agricultural equipment and
    domestic heating and refrigeration equipment,
    auxiliary machinery in power stations, conveyor
    belts, lifts, or machine tools.
	8 000–15 000

Machines for use 8 hours per day and not always fully
    utilized:
    general purpose gear units, stationary electric motors.
	10 000–25 000

Machines for use 8 hours per day and fully utilized:
    machine tools, wood processing machinery,
    machines for the engineering industry, cranes
    for bulk materials, ventilating fans.
	20 000–30 000

Machines for continuous use 24 hours per day:
    compressors, pumps, fans, associated stationary
    electric machines and mine hoists, marine
    propulsion equipment, ships propeller shaft
    thrust bearings, tunnel shaft bearings.
	25 000–50 000

Machines where high reliability is required:
    waterworks machinery, marine pumps, printing
    machinery, single stream pumps.
	100 000

* $L_{10(h)}$ is related to mean time between failures by:

$L_{10(h)} = 0.1 \times$ mean time between failures (hours) $\times$ number of bearings.

Note that the nominal life values here are in hours. The designer should calculate the corresponding life in millions of revolutions using the equation

$L_{10} = L_{10(h)} \times N(\text{rev/min}) \times 6 \times 10^{-5}$

tion of life adjustment factors into L_{10} life equations to take these into account.

ADJUSTED BEARING LIFE

$L_{na} = a_1 a_2 a_3 . L_{10}$

a_1 = life adjustment factor for reliability
a_2 = life adjustment factor for material
a_3 = life adjustment factor for operating conditions

The life adjustment factor for reliability, a_1, can be used to design for greater reliability than 90 percent, appropriate values being given in the bearing manufacturers' catalogues for 95–99 percent, giving L_5 to L_1 lives. Factors a_2 and a_3 are interdependent and are usually combined as an a_{23} factor. The improved material properties have been taken into account in the catalogues. For normal conditions where elasto-hydrodynamic lubrication occurs, a value of 1 can be taken for a_{23} under heavily-loaded conditions; where there is a breakdown in elasto-hydrodynamic lubrication, reduced values have to be taken, and lubricants containing extreme pressure additives have to be used to obtain reasonable lives. These conditions are described more fully when we come to consider the lubrication of rolling bearings.

It should be noted that the calculated L_{10} life is very sensitive to the load and is inversely proportional to the power 3 in the case of ball bearings, to the power 10/3 for roller bearings. In many industrial machines the load is not accurately known, for example, hydraulic loads arising in centrifugal pumps, loads arising from belt tension in belt drives, and magnetic unbalance loads in electric motors. This uncertainty has to be borne in mind when basing designs on the valuable practical experience embodied in Table 6.1; it is reflected in the wide ranges shown in this table.

6.2 SELECTION OF ROLLING BEARINGS

Rolling bearings are available in a large variety of types and configurations. Table 6.2 summarizes the characteristics of the main types found in industrial applications. Many of the bearings are available in different dimension series, that is different outer diameters and widths for the same inner diameter. Alphanumeric codes are used to define the different types of dimension series (ISO 15 for radial bearings, except for taper roller bearings which are covered by ISO 355).

Although it is easy to check from the manufacturers' catalogues whether a bearing will meet the required duty, there is no simple way of selecting the optimum bearing for any particular application. (If required, a more rigorous method of selection is available in Engineering Sciences Data Unit, Data Items 81005, 81037, 82014.) The guiding principle should be to use the simplest bearing arrangement that will meet the duty. (Table 6.2 has been arranged in rough order of increasing complexity.) Some bearings are available with additional features, for example: deep groove ball bearings can be obtained with integral metal shields (suffix Z) and rubber seals (suffix RS), pre-packed with grease; ball bearings can be provided with filling slots to allow the insertion of additional balls, giving greater load capacity, though of course limited to one direction of thrust; snap ring grooves (suffix N) can be machined in the outer race to give axial location in the housing. In general, such special types are less readily available, have limitations, and should be avoided except in special circumstances. The points outlined below should be borne in mind when making an initial selection.

SELECTION OF BEARING TYPE

- Simplest bearing arrangement gives the greatest reliability.
- Simple types of bearings are easier to fit and lubricate.
- Simple types of bearings are the cheapest and most readily available as spares.
- Use smallest size consistent with adequate load-carrying capacity.

The simplest arrangement is a deep groove ball bearing at one end of the shaft and cylindrical roller bearing at the other. This combination of a 'locating' bearing at one end and a 'non-locating' bearing at the other obviates any risk of overloading the bearings through axial expansion of

Table 6.2   Principle types of rolling bearings for industrial use

Type	ISO Code[1]	Characteristics
Deep groove ball bearing	6 xxx	The basic type for radial loads with some capacity for axial loads
Cylindrical roller bearing	NXX xxx[2]	The basic type for radial loads; special designs with additional flanges can carry some axial load
Single row angular contact ball bearing	7 xxx X[3]	Enhanced axial load capacity in one direction compared with deep groove ball bearing, normally used in opposed pairs
Double row spherical roller bearing	2 xxxx	Combined high radial and axial load capacity

Nu   N   NT   NuP

Contact angle   Face-to-face   Back-to-back

Taper roller bearing	3 xxx	High axial load capacity, but capable of carrying radial loads, frequently used in opposed pairs
Four point contact ball bearing	QJ xxx	Limited radial load capacity, but suitable for both directions of axial load (inner race split)
Double-row angular contact ball bearing	3 xxx	A Siamesed pair of angular contact ball bearings, giving both radial and axial load capacity
Needle roller bearing	(see Note 4)	High radial load capacity, used when limited radial space available, can be run directly on hardened shaft

Notes

1 For the purpose of illustration the codes shown in this table are divided into two parts; the first part defines the type, the second the dimensions. In normal practice the code appears without a space. X = a letter, x = a digit.
The last two digits of the second part of the code give the bearing bore: 00 = 10 mm, 01 = 12 mm, 02 = 15 mm, 03 = 17 mm, for larger sizes the figures give the bore in mm divided by five.
The third digit from the end gives the diameter series,
e.g., 2 = light series, 3 = medium series, 4 = heavy series.
The fourth digit gives the width series, ranging from 0 to 6 (increasing number indicating increased width for the same outer diameter); digit 0 is normally omitted.

2 The initial part of the code may consist of 1–3 letters that define the disposition of the flanges on the races.

3 The suffix code defines the contact angle: C = 15, E = 25, A = 30, B = 40.

4 Coding is complex: see manufacturers' catalogues.

the shaft through differential temperature between rotor and stator. Where this arrangement does not provide sufficient axial load capacity, the deep groove bearing can be replaced by a pair of angular contact bearings or a double row spherical roller bearing.

6.2.1   Internal bearing clearance

In order to operate satisfactorily, the rolling elements must be able to roll freely between the races without jamming. To avoid jamming and skidding of the rolling elements with consequent failure of the bearing, bearings are manufactured with internal clearance. The greater the internal clearance, the less accurate the location of the shaft, and the greater the noise during running. Mounting the races with an interference fit and thermal gradients across the bearing during running will affect the internal clearance. In order to obtain a suitable compromise between these factors and the risk of loss of clearance during running, rolling bearings are manufactured with different ranges of internal clearance to suit different applications so that optimum clearance is obtained under running conditions.

Details of the internal clearance class codings are given in Table 6.3 together with a rough indication of the radial temperature gradient that can be accommodated. The different clearance classes are obtained by selective assembly, with 'normal' clearance most readily available.

The actual value of internal clearance provided is a function of bearing size. The clearances of different types of bearings are not identical; for example, a cylindrical roller bearing will have a greater clearance than a deep groove bearing of the same size and clearance designation. The clearance values for deep groove ball, cylindrical roller, and double row spherical roller bearings are standardized internationally.

Table 6.3   Internal clearance in rolling bearings

Code	Description	Approximate allowable temperature gradient
CI ⎫	Below normal	
C2 ⎬	clearances	
CN[1]	Normal clearance	10°C
C3 ⎫	Above	25°C
C4 ⎬	normal	40°C
C5 ⎭	clearance	

[1] CN is not normally marked on the bearing

6.2.2 Selection of internal clearance class
Selection of the correct clearance class depends on the following factors.

FACTORS INFLUENCING CHOICE OF CLEARANCE CLASS

- The method of fitting
- The thermal gradients across the bearing in service

The usual method of mounting is to fit one race with an interference fit (normally the race that is rotating with respect to the load line) and the other race with a push fit. The interference fit results in some loss of clearance. This has been allowed for in 'normal' clearance bearings. In the more unusual case when both races are given interference fits, a high clearance bearing, e.g., C3, should be used.

Where there is heat conducted along a shaft to the bearing the differential expansion across the bearing results in some loss of clearance. To compensate for this higher than normal clearance bearings are used. For example, C3 bearings are used on larger electric motors, fans handling hot gases, and pumps handling hot liquids. At high speeds there is a tendency for increasing thermal gradients to be set up across the bearing; when operating at speeds above eighty percent of the manufacturer's rated maximum speed for a bearing, a C3 bearing is advisable.

Bearings can run without internal clearance, elastic deformation in the rolling elements and races preventing jamming. This is, however, a somewhat sensitive condition and is only used where very precise shaft location is required.

No additional risk of failure is introduced by using higher clearance bearings, so in cases of doubt C3 bearings should be used. It is usually not difficult to obtain C3 bearings, but the availability of the other clearance classes is very restricted.

6.2.3 Bearing fitting
Recommended fits for shaft seatings and bearing housings are given in the manufacturers' catalogues. The tolerance limits, however, are very restrictive and, while acceptable in equipment manufacturers' assembly shops, are impracticable for fitting replacements in local workshops. This prob-

lem is compounded by the fact that different bearing manufacturers tend to recommend different fits. As the bearings are manufactured to very restrictive standards, the justification for this is somewhat doubtful. The practical approach is to work within the range given by the different manufacturers: this not only eliminates any difficulty in using replacement bearings from different manufacturers, but also provides some allowance for the inevitable loss of material that occurs with repeated replacement.

The full bearing code is normally stamped on the side of one of the races (though clearance group CN is normally omitted).

6.3   LUBRICATION OF ROLLING BEARINGS

There are two requirements for lubrication in rolling bearings.

LUBRICATION REQUIREMENTS OF ROLLING BEARINGS

  – Sliding contacts at the cage
  – Rolling contacts between rolling elements and races

In addition, the lubricant should have good corrosion-protecting properties to keep the finely finished steel contact areas free from corrosion.

6.3.1 Method of lubrication

Oils and greases are used as lubricants. One of the great advantages of rolling bearings is the simplicity of lubrication with grease. At low speeds the bearing can be fully packed with grease so that the grease can act both as a lubricant and a seal against dirt and moisture. At higher speeds the bearing cannot be fully packed with grease as the heat generated by viscous shear will cause it to degrade. The correct method is to fill the space between the races and rolling elements fully with grease, but to allow sufficient space in the end covers to allow the excess to be thrown out from the bearing when it is running. Lubrication is carried out by the thin film of lubricant that remains on the races; the grease in the housing acts as a seal and provides a reservoir of oil to keep the bearing lubricated. When a correctly packed bearing is started the temperature rises until the grease is cleared from the bearing and then falls to its equilibrium running value:

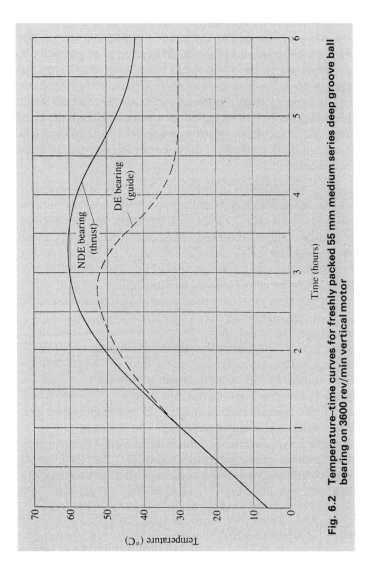

Fig. 6.2 Temperature–time curves for freshly packed 55 mm medium series deep groove ball bearing on 3600 rev/min vertical motor

this process may take several hours (Fig. 6.2). In many cases it is possible to run a grease-packed bearing for periods of up to two or more years. Eventually, however, the grease deteriorates and either fresh additions have to be made or the grease changed. Deterioration of grease is a function of temperature and the rate of shear. It is also affected by the type of bearing (grease life is lower in double row bearings and taper roller bearings) and the type of mounting (grease life is reduced in vertically-mounted bearings because of increased shear effects through the tendency of the grease to slump back into the bearing). The following method allows an estimation to be made of the relubrication interval for grease in grease-packed bearings.

(1) Using Fig. 6.3 obtain the nominal maximum speed for grease lubrication for the particular size and series of bearing. (This plot is based on the maximum speed for grease lubrication for deep groove ball and cylindrical roller bearings taken from the bearing manufacturers' catalogues.)

(2) Calculate the specific maximum speed for grease lubrication for the application being considered by multiplying the nominal maximum speed by the following factor(s) to allow for the effect of different bearing designs and mounting arrangements:

    bearings on vertical shafts   0.75

    bearings mounted in adjacent pairs without spacers   0.75

    bearings with cages centred on inner races   0.70

    taper and double row spherical roller bearings   0.35

    bearings with rotating outer races   0.35

(3) Determine the speed rating, that is the actual running speed as a percentage of the specific maximum speed.

(4) Determine the grease relubrication interval from Fig. 6.4. (This plot is derived from practical experience on operating machines and is based on a lithium soap grease operating at a maximum temperature of 70°C.)

Experience with continuously running plant suggests the practice for grease lubrication given in Table 6.4.

If regreasing during running is required then it is necessary to provide some positive means of discharging the excess grease so that the bearing does not become overfilled. This is done by means of a grease relief valve,

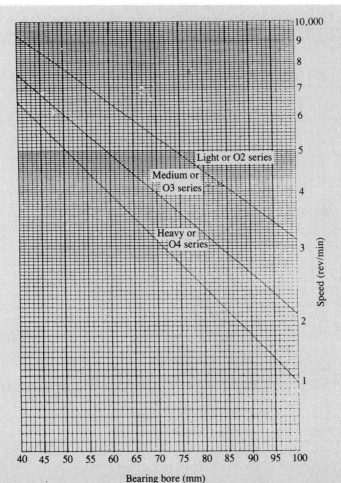

Bearing bore (mm)

Fig. 6.3 Maximum speed for grease lubrication for horizontally mounted deep-groove ball cylindrical roller bearings with rolling element centred cages

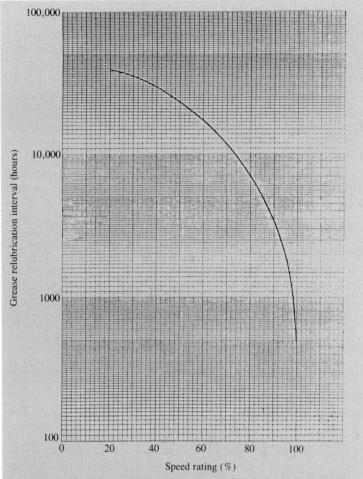

Speed rating (%)

Fig. 6.4 Relubrication interval for lithium soap grease as a function of bearing speed rating

Table 6.4 Recommendation for grease lubrication of rolling bearings

Calculated grease life (h)	Method of lubrication
>15 000	Grease pack
<15 000 >1 000	} Relubricate 4-weekly
<1 000	Not suitable for grease lubrication

which consists of a rotating flinger on the shaft that ejects any grease coming into contact with it (Fig. 6.5). If the calculated life is less than 1,000 h, regreasing is no longer economic and oil lubrication should be used.

A more elaborate method of determining grease life, using additional empirical data and taking additional factors into account, is available in Engineering Sciences Data Unit, Data Item 78032.

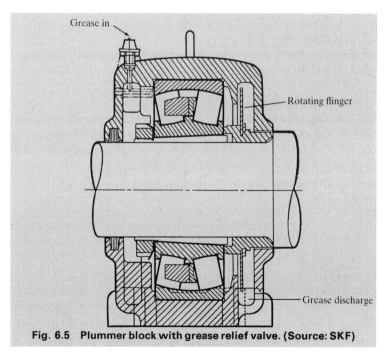

Grease in

Rotating flinger

Grease discharge

Fig. 6.5 Plummer block with grease relief valve. (Source: SKF)

Lubrication with oil is relatively simple, though more sophisticated sealing methods, e.g., rubber lip seals, are required. Oil lubrication can be achieved by a bath, the level being controlled to about half way up the lowest ball or roller, by splash using a ring or disc oiler, or by circulation – the method being chosen to limit the oil temperature in the bearing to about 70°C. Oil-mist systems, where the oil is transported to the bearing as an aerosol in a pressurized stream of air, are useful when a large number of bearings have to be lubricated. It can either be used as a 'purge system', where the oil is fed to a bearing pedestal, topping up the sump, or as a 'pure mist' system, where it is the only source of lubricant. Purge systems are useful in maintaining a positive pressure in the bearing housing that helps to prevent the ingress of contaminants. Pure mist systems require more sophisticated engineering to guard against failure of supply.

6.3.2 Selection of lubricants

Lithium greases are suitable for most applications where bearing temperatures do not exceed 70°C; lithium complex greases allow operation up to 100°C (see Table 3.4).

Oil type is not critical provided an elasto-hydrodynamic lubricating film is generated at the rolling contacts. From a practical point of view this is achieved if the viscosity at the bearing temperature is 12 cP for ball and cylindrical roller bearings, 20 cP for taper roller and spherical roller bearings. (This applies also to the oil component of the grease.) Where this cannot be achieved a lubricant containing an ep additive should be used. Table 3.3 gives recommendations for where ep lubricants should be used. Only in critical cases does the specific film thickness have to be calculated; in the majority of applications the load rating of the bearing, C/P, gives an adequate guide.

6.4 BEARING 'EQUIVALENTS'

Care has to be taken before replacing bearings with an 'equivalent' from a different manufacturer.

Particular points to which attention should be paid are cage types for grease lubricated bearings, contact angle with angular contact bearings (manufacturers use different contact angle bearings for their standard

BEARING EQUIVALENTS

'Equivalent', as normally used in connection with rolling bearings, refers only to external dimensions and does not necessarily imply functional equivalence.

supply), and the use of filling slots to give increased radial load capacity. A bearing with a pressed steel cage allows easier access of grease than one with a machined cage and is better for bearings operating near the limit for grease; plastic cages do not perform as well as metal cages in grease-lubricated angular contact bearings. A bearing with a contact angle of 20 degrees may fail catastrophically if it is used to replace one with a contact angle of 40 degrees chosen to give high axial load capacity. For example, in pumps where significant axial loads have to be carried it is safer to standardize on 40 degree contact angle bearings.

6.5   ROLLING BEARING FAILURES

Rolling bearings may fail by a number of mechanisms. Some knowledge of these is desirable in order to distinguish those that have occurred at the end of the design life from those caused by bad fitting, maloperation, lubrication breakdown, and so on.

ROLLING BEARING FAILURE MECHANISMS

Fatigue	Normal end of design life
Premature fatigue	Result of damage while stationary:
	– false brinnelling;
	– contact corrosion;
	– fitting damage;
Lubrication failure	Cage wear
	Breakdown of elastohydrodynamic lubricating film, giving surface distress
Wear	Contamination by solids
Plastic deformation	Loss of internal clearance

Fig. 6.6 Fatigue failure of inner race of taper roller bearing

6.5.1 Fatigue

As mentioned above, rolling bearings ultimately fail by fatigue. This can occur either on the rolling elements or on one of the raceways. It starts from a single pit and proceeds from this point (Fig. 6.6). Practical experience, however, suggests that very few bearings actually reach their fatigue life in service. If early failures occur it is worth examining the design to see if the initial assumptions about the load were correct.

6.5.2 Premature fatigue

Any damage to the surface reduces the fatigue strength and sets up a point of weakness from which fatigue pitting develops. Three cases predominate in practice.

(1) *False brinnelling.* If rolling bearings are subject to vibration when stationary, indentations develop at the points of contact (Fig. 6.7). To prevent damage, machines subject to vibration should be rotated about half a turn once per week. It should be recognized that false

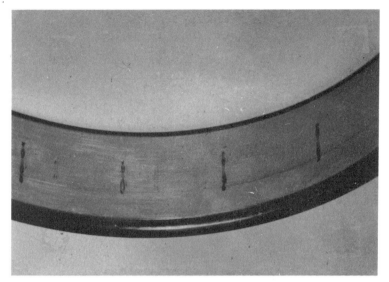

Fig. 6.7   False brinnelling of cylindrical roller bearing

brinnelling can occur even before bearings are fitted if they are subjected to vibration in storage.

(2) *Corrosion*. New bearings are packed with corrosion preventives.
However, if they are cleaned and packed with grease they become
liable to corrosion unless rotated so that all the surfaces are coated
with grease. If this is not done, corrosion can occur at the contacts
between rolling elements and races (Fig. 6.8).

(3) *Bad fitting*. If loads are transmitted through the rolling elements
during fitting, i.e., by fitting an inner race on to a shaft by applying
load through the outer race, indentations are caused on the raceways. Cocking a shaft with respect to the housing when fitting a
cylindrical roller bearing can cause score marks on the raceway
(Fig. 6.9).

The common feature in premature fatigue is that patches of fatigue
develop at rolling element spacing (Fig. 6.10), allowing it to be readily distinguished from normal fatigue.

Fig. 6.8   Corrosion damage on cylindrical roller bearing

6.5.3   Lubrication failure

Under normal operating conditions, loading at the cage contacts is low and satisfactory lubrication can be easily provided. Cage wear occurs, however, with grease lubrication if the grease is used beyond its life (see above), or when the contact loads are increased by operating the bearing with misalignment (Fig. 6.11).

Breakdown of the elasto-hydrodynamic lubrication film at the rolling contacts results in a roughening or scuffing of the surfaces (Fig. 6.12), usually described as surface distress.

6.5.4   Contamination

Contamination of the lubricant by liquids can cause corrosion of steel parts, including the cage.

Solids cause wear. If the solid particles are very fine, the tracks take on a specular, burnished appearance: coarse particles give rise to dull, matt tracks on the races (Fig. 6.13). Wear results in increased bearing clearance, but does not of itself usually give rise to failure.

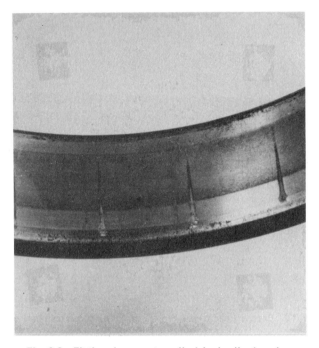

Fig. 6.9   Fitting damage to cylindrical roller bearing

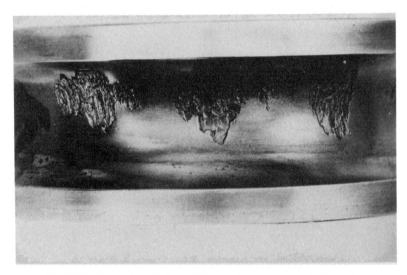

Fig. 6.10   Premature fatigue failure of deep groove ball bearing

Fig. 6.11   Cage wear of rolling bearing caused by grease lubrication failure

Fig. 6.12 Scuffing of rolling bearing race caused by breakdown of elasto-hydrodynamic oil in film

Fig. 6.13   Wear of roller bearing track caused by particle contamination

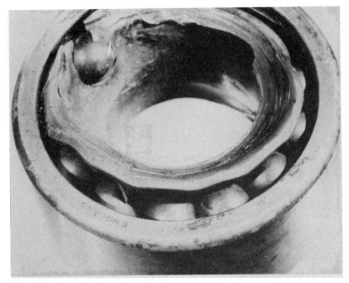

Fig. 6.14   Plastic deformation of ball bearing. Note progressive damage from
outer to inner race

6.5.5   Plastic deformation failure

General plastic deformation usually indicates gross overloading: typically, the damage decreases from inner to outer race, where the housing provides a good heat sink and reduces the temperature (Fig. 6.14). Most frequently this arises from loss of internal clearance caused by excessive interference fits on the shaft or housing, or high thermal gradients across the bearing. Paradoxically it can also follow from lack of interference fit of the race subject to rotating load. This allows the race to creep on its seating; this causes wear and ultimately the race starts to rotate and then expands because of the frictional heat generated taking up the internal clearance. Loss of clearance can also result from excessive heat generation resulting from a complete breakdown in lubrication.

CHAPTER 7

Methods of Lubrication

Methods of lubrication: *self-contained systems* versus *circulation systems*
  - choice primarily determined by *cooling requirement*
  - removal of heat generated by friction

Circulation system designed to ensure separation of contaminants and prevention of foaming

Monitoring aimed to protect machines not circulation system.
Gears and *reciprocating compressors* impose particular requirements.

Having considered the fundamental principles involved in hydrodynamic lubrication and the materials used as lubricants, some methods of applying the lubricant to different mechanisms and ways of selecting the appropriate viscosity grade will now be discussed.

7.1 PLAIN BEARINGS

It is possible to lubricate plain bearings by intermittent oiling, or by drip and wick feeders. However, such methods, with their high labour cost and unreliability, cannot seriously be considered for modern industrial machinery. At slow speeds where heat generation is not a problem, self-lubricating materials or self-contained systems, such as oil-impregnated porous metal bearings or oil and grease impregnated plastics, should be used rather than plain bearings that require a regular oil feed. Lubricated mechanisms require only a small amount of oil to provide a hydrodynamic film. Except for slow speeds, it is necessary to supply a flow of oil to remove the frictional heat generated at the contacts; lubrication system design is dictated by the cooling flow requirements. In order to calculate the required oil flow to control the temperature at the contact it is necessary to use an iterative method to solve the interdependent hydrodynamic and heat balance equations. Methods for this are beyond the scope of this book; the treatment here is limited to practical considerations that determine satisfactory performance.

7.1.1   Self-contained systems: ring and disc oilers

The simplest type of self-contained oil lubrication system for the plain bearing is the sump and ring oiler (Fig. 7.1). The oiling ring is a simple and very effective method of providing a regular circulating supply of oil to a bearing. The design of ring-oiled bearings presents formidable theoretical difficulties as the oil supply rate depends on the viscosity in the sump – a function of the heat balance in the system. Designs of ring-oilers are largely based on experience and it is thus worth examining their practical limitations. Oil is carried from the sump on the inner periphery of the ring and is transferred to the shaft at the ring contact. At low speeds there is no slip between the ring and the shaft; oil flow is directly proportional to the rotational speed of the ring. With increasing speed a hydrodynamic film is generated between the ring and shaft and the frictional drag on the ring of the oil in the sump causes it to slip so that the rate of oil delivery no longer increases in direct proportion to the speed. At still higher speeds centrifugal force throws the oil off the ring and it is no longer available to lubricate the bearing, though even before this stage is reached the oil delivery may be inadequate to satisfy the cooling requirements of the bearing. The rate of oil delivery is also a function of the viscosity of the oil. The situation is further complicated by the heat generated by viscous loss in the bearing that increases the temperature of the oil in the sump and decreases

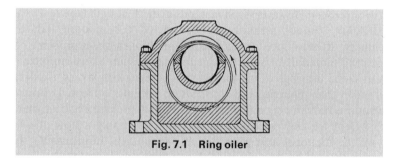

Fig. 7.1   Ring oiler

its viscosity, the equilibrium temperature being determined by the heat losses from the bearing pedestal. It is this heat balance that determines the maximum speed at which satisfactory lubrication can be guaranteed. The limit is raised if external cooling is applied to the oil in the sump.

In order to transfer oil to the bearing, the inner surface of the ring has to be wetted. Any unnecessary immersion, however, only increases the drag on the ring and reduces the oil delivery; depending on ring size, 10–15 mm gives a good compromise between reliability of wetting and excessive immersion. Oil level in the sump is obviously critical to ensure the right degree of wetting and is best controlled by a constant level oiler (Fig. 7.2).

The ring dimensions and material do not appear to be critical as long as the ring is sufficiently heavy to provide sufficient friction against the journal to overcome the drag of the oil in the sump. Brass is suitable and plain rectangular section rings appear to give the best results. Continuous rather than jointed rings are preferred; the latter tending to snag and stop rotating unless the joints are made very carefully. The following proportions are typical of good practice – ring internal diameter: journal diameter 1.5; ring internal diameter:ring width 10; ring width: ring thickness 3. The practice of machining circumferential grooves on the inner periphery of the ring in order to increase the amount of oil picked up is not recommended since at high speeds the oil is merely flung to the bottom of the groove and does not wet the shaft; moreover, the increased loading of the ring can sometimes lead to wear, with the ring cutting a groove in the shaft.

Older practice was to use wide bearings with two oiling rings: this yields no advantage and current designs are based on single rings. With good designs ring oilers can be used up to peripheral journal speeds of 10 m/s when there is no provision for additional cooling. With water cooling of the oil in the pedestal satisfactory lubrication can be achieved at speeds up to 20 m/s: VG46 and VG68 oils give the best results.

Disc oilers, which rely on a disc attached to the shaft dipping into the oil in the sump for oil transfer, overcome the problem of slippage of a loose ring (Fig. 7.3). The disc is made with a 'U' section to increase the pick-up of oil and the oil is transferred by means of a plastic scraper to an upper sump where it can be cooled by a heat exchanger before it flows by gravity to the journal. While apparently giving more certain delivery of oil than a loose ring, disc oilers are more complex and require more care in setting up and in oil level control. During operation, part of the oil is transferred to

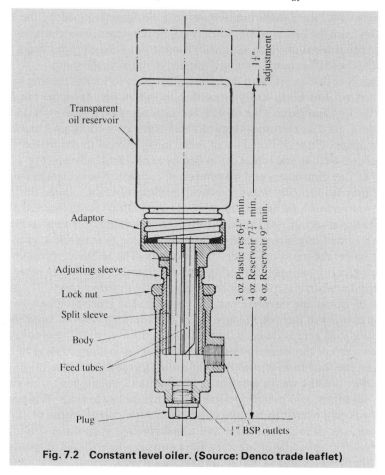

Transparent
oil reservoir

Adaptor

Adjusting sleeve

Lock nut

Split sleeve

Body

Feed tubes

Plug

$1\frac{1}{4}''$ adjustment

3 oz Plastic res $6\frac{3}{4}''$ min.
4 oz Reservoir $7\frac{3}{4}''$ min.
8 oz Reservoir $9''$ min.

$\frac{1}{4}''$ BSP outlets

Fig. 7.2   Constant level oiler. (Source: Denco trade leaflet)

the upper sump, so that two levels have to be marked for the bottom sump – a stationary one and a running one; the risk of confusion is obvious. It is doubtful with this additional complexity, if they offer much advantage over the well-designed ring oiler; maximum speeds for satisfactory operation are still limited to shaft peripheral speeds of 20 m/s. Best results are given with a VG32 oil.

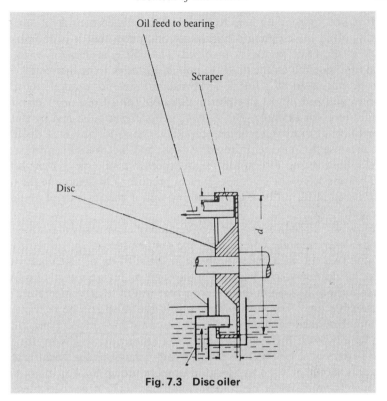

Fig. 7.3   Disc oiler

7.2   OIL CIRCULATION SYSTEMS

Circulation systems are necessary for higher speeds than can be satisfactorily lubricated with self-contained systems and also where there are additional items requiring lubrication, e.g., gears, flexible couplings, etc. The function of a lubrication system is to reliably deliver the required quantity of clean lubricant to the various parts requiring lubrication. Basically this means a separate reservoir, oil pump, and cooler, together with some means of maintaining the lubricant in a clean condition. The requirement of reliability introduces a degree of complexity into the system, with duplication of pump, filter, and cooler.

It is not appropriate here to go into the detailed considerations that are involved in designing a circulating oil system, but it is desirable to look at the basic principles involved. The first consideration is the oil feed rate required by the different points that have to be lubricated. This can be calculated by a full design procedure (see, for example the Data Items published by the Engineering Sciences Data Unit) or by empirical guides (see, for example, *The Tribology Handbook* published by Butterworth). In determining pumping capacity, allowance has to be made for the increased requirement caused by wear and increase of bearing clearances, limits being determined by an acceptable amount of wear before the bearing has to be changed. Typical practice is to determine the total rate required for all bearing points and add a contingency of about 25 percent.

Flow rate through the system is a function of oil viscosity and it is necessary to exercise some control over the temperature of the oil fed to the pump. This can be a major problem in the start-up of machines with a large lubrication system installed in the open, and it is essential to provide some form of heating in the reservoir that will raise the temperature in a reasonable time. The power input to the heater should be restricted to ensure that cracking of the oil does not occur at the surface; typical limiting figures where there is only convection heating are 15 kW/m^2 for electric heaters and 3.5 MN/m^2, $160°C$ steam conditions for steam heaters. Heaters should be sized to bring the temperature up to $20°C$ in not more than eight hours under the coldest conditions likely to be encountered at start-up.

The oil rates to the different parts are best controlled by means of orifices. This means that a major leak at one lubrication point does not starve all the others, so that damage can be limited.

7.2.1 Maintenance of oil quality

Oil quality depends on three things: the absence of adventitious contaminants; the degradation of the oil by oxidation or additive depletion; and the complete separation of gas from the oil in the reservoir so that it is not pumped round the system.

It is convenient to consider the mechanisms associated with these separately and the provisions that have to be made to combat them, though one measure that is common to all is the reservoir design. Provision of baffles to separate the oil return from the pump suction does much to assist breakdown of foam and allow contaminants to settle.

The reservoir should have a sloping bottom and drain to allow separated contaminants to be removed, but the most important factor is relative stillness and time in the reservoir. This is best expressed as the residence time, defined as the oil capacity divided by the pumping rate. The residence time required is a function of the degree of contamination; where there is a steady ingress of contaminants because of the nature of the environment, e.g., water contamination of steel rolling mill lubrication systems, residence times of up to an hour may be necessary. This, of course, means a very large reservoir and for more normal applications where the main problem is the separation of air bubbles entrained at the bearings residence times will be less than 10 minutes. Values of 8–10 minutes are preferred with oil systems for rotary machines.

7.2.2 Contamination

Solid and liquid contaminants are clearly undesirable in lubrication systems and provision has to be made to eliminate them as far as possible. In bearings and seals, film thicknesses may be as low as 1–2 microns, but it is clearly impracticable to filter the large flow rates required to this degree. In most industrial systems the rate of contamination by external solids during running is low, the major 'dirt' problem arising either when the machine is first brought into commission or after major overhauls. Once the bulk of the dirt has been removed a 25 micron filter is adequate for bearings and gears, a 5–10 micron filter for seals and turbine control systems, particularly in conjunction with a by-pass cleaning system handling about 10 percent of the system capacity per hour. By-pass cleaning systems can be based on filters or centrifuges to remove solids and coalescers or centrifuges to remove immiscible liquids. In order for a centrifuge to remove water adequately the oil to the centrifuge has to be heated to 80°C. For high pressure hydraulic systems, where the oil flow rate is much lower, 5 micron filters need to be used.

Water contamination is frequently more of a problem with industrial machines: this can occur from atmospheric breathing, steam leaks from turbine drivers, and water leaks from cooling systems. Water separates readily from oil, though oil oxidation products tend to slow down the rate of separation. Separation is further inhibited by the presence of iron corrosion products, so that if the water is not regularly removed stable emulsions can be formed that resist breakdown even by centrifuge. This illustrates the importance of adequate residence time in the reservoir to allow separation and the need for by-pass centrifuges or coalescer-filters when residence times are low or the rate of water contamination is high. For low rates of contamination and in buildings where the atmospheric humidity is low, an extractor fan drawing air over the surface of the hot oil in the reservoir provides an extremely effective way of removing water. This is a good reason for maintaining the temperature in the reservoir at about 60°C so that the vapour pressure is high enough for the water to be removed. Reduction of internal corrosion is effective in controlling emulsion formation. This applies particularly to return oil pipes and reservoirs where the surfaces are not completely protected by oil; these should be of corrosion-resistant materials or suitably protected against corrosion. Internal corrosion will of course also give an increased rate of solid contamination of the oil and should be avoided for this reason alone.

7.2.3 Lubricant degradation

Oil degradation, in addition to stabilizing emulsions as already indicated, results in an increase in the oil viscosity and is eventually responsible, if allowed to proceed far enough, for the precipitation of sludge. Oil degradation products also reduce the rate of air separation and foam breakdown (see below). Degradation occurs either by oxidation or by thermal cracking of the oil; both are a function of temperature, the former depending, in addition, on the presence of oxygen. Mineral oils in the absence of oxygen do not degrade until temperatures are above 300°C (this is why they can be used successfully as heat transfer media in enclosed systems), and hence thermal degradation is only a problem at the surface of heaters when the heat flux is too high. Oxidation occurs above 50°C, the rate increasing with increasing temperature, and is strongly affected by the presence of catalysts. Temperatures of about 120°C can be permitted in bearings because the access of oxygen is low and the time at temperature very short, but in return lines and reservoirs 60°C is a reasonable maximum. As already mentioned there is an advantage in keeping oil temperatures in reservoirs reasonably close to this limit as it assists in the removal of water by evaporation.

7.2.4   Air entrainment and foaming

It has to be recognized that intimate mixing of air and oil occurs at all high-speed lubricated contacts. Lubrication systems have to be designed to minimize air entrainment as well as to provide ample opportunity for detrainment so that bubbly oil is not pumped round the system.

Orifices in lubricant spray pipes for gears should be smooth to reduce the risk of pick-up of air by the jet of oil. With high-speed gears large quantities of oil are required for cooling. Spraying the gear teeth at the exit from the nip helps to minimize aeration as well as reducing heat generation through unnecessary viscous losses. Oil return lines should run only half-full to allow the maximum opportunity for bubble separation, and the drain oil should be returned to the reservoir at or about the oil level to reduce the risk of splashing. A wire gauze foam arrester (about 100 mesh) in the first compartment of the oil reservoir fitted completely below the oil level at an angle of about 45 degrees, helps to prevent the flow of bubbles to the pump suction compartment. Residence times in the reservoir should be related to the rate of air entrainment in the system. When this is high, as for instance with high-speed bearings, gears, or lubricated gear couplings, a minimum of eight minutes residence time should be the aim. Again, maintenance of the oil in the reservoir at about 60°C, already recommended for the removal of water, is helpful in speeding up air detrainment.

7.2.5   Warning and protective devices

The final aspect of importance with lubrication systems concerns warning and protective devices. This area covers warning devices that indicate when remedial action is required; protective devices that take action against component failure or trip the machine; and monitoring devices that follow the behaviour of the lubricated parts and allow action to be taken to reduce the risk of consequential damage.

Oil pumps, filters, and coolers on large, important systems are usually duplicated. It is normal practice to cover main oil-pump failure by switching in the spare or auxiliary pump on low system pressure and to protect against loss of oil from the system by a low level indicator in the reservoir. Machine trips are provided at lower values of pressure or oil level in the event of the remedial action not being successful, the trip having a five second delay to avoid spurious trips on momentary fall of pressure when switching in the spare pump. Typical values are:

OIL SYSTEM PRESSURE MONITORING

Normal supply pressure	2.5 bar (0.25 MN/m²)
Alarm and switch in of spare pump	1.9 bar (0.19 MN/m²)
Trip	1.3 bar (0.13 MN/m²)

This, however, does not give adequate protection in the case of high-speed machines with long run-down times. In the event of a major leak, switching in of the auxiliary pump merely empties the system more quickly and there is no lubricant available to cover the run-down. Again, in the case of systems in which the main oil pump is driven from the machine, the auxiliary pump has to take over during the shut down when the delivery from the main pump begins to fall off; if the auxiliary pump fails to start no oil will be available for the shut down. This situation can be covered by installing a gravity or pressurized head tank that is fed from the pump delivery and is of sufficient capacity to discharge through the bearings during the run-down period (Fig. 7.4). It has to be remembered that the main function of a circulation system is to remove heat from the bearings, even after the machine has come to a stop. Some turbine and compressor rotors operate with temperatures of 350°C and above. Oil flow must be maintained to conduct this heat away and prevent melting of the white metal bearings.

Other conventionally fitted instruments on lubrication systems are differential pressure gauges across filters to indicate the time for replacement, and differential temperature indicators across coolers to indicate the need for cleaning.

In general, however, the provisions for indication of incipient bearing failure are not sufficiently sensitive for remedial action to be taken in time and there is a need to look at the philosophy of protecting machines against the consequences of tribological failure, recognizing that such failures may occur on large machines. It is one of the advantages of the low strength and low melting white metal alloys used in bearings that the bearing fails first and, provided sufficiently quick action is taken, damage can be limited to the comparatively cheap and more easily replaced bearing. In the case of steel components, rolling bearings, and gears, failures tend to be more gradual and more time is available for action to be taken before there is complete collapse.

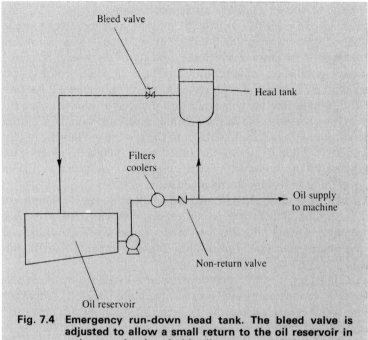

Fig. 7.4 Emergency run-down head tank. The bleed valve is adjusted to allow a small return to the oil reservoir in order to maintain suitable oil temperature in head tank

In principle, the early stages of component failure can be detected by one of the following mechanisms:

POSSIBLE TECHNIQUES OF MONITORING
COMPONENT DEGRADATION

- an increase in the rate of generation of wear particles
- a change in the energy loss in the bearing
- an increase in the freedom of movement at the bearing contact
- an increase in noise and/or vibration

Some use has been made of routine spectroscopic analysis of lubricants from enclosed systems to provide an indication of increased wear rates (SOAP – spectroscopic oil analysis programme). This is obviously restricted to those cases in which the capacity of the system is limited so that there is sufficient concentration to detect the wear products and, in addition, there is ample time for remedial action to be taken. It has proved useful in monitoring the behaviour of a number of similar systems, like railway diesel engines, but has not been as successful with large industrial systems, though it is a useful adjunct to the routine monitoring of oils, as described in Chapter 4. On the other hand, magnetic plugs suitably positioned in the system are extremely effective at detecting the early stages of failure of rolling bearings and gears.

An increase in energy loss means an increase in power consumption and heat generation. The former is not usually sufficiently sensitive, though motor power consumption has been used to indicate roller bearing failures in large rotary machines. Temperature rise does provide a useful indication.

It is important, however, that the detector is located at the point of heat generation. In most lubricated contacts the bulk of the oil is supplied for cooling and hence a detector in the lubricant drain gives a much slower warning than a detector at the solid surfaces, e.g., in contact with the back of the white metal in thrust pads or in the loaded region of a journal bearing. These give a rapid response and can be effective in preventing damage of the machine by operating alarms or trips on a temperature rise of as little as 5–10°C. Equally, a sudden and unexplained drop in bearing temperature should also be taken as a warning. This can happen if the white metal is completely wiped out, the journal continuing to run on the steel backing, but the increase in oil flow rate through the bearing effectively reducing the temperature. Machines may be able to run in this condition for short periods, but it is clearly an unsatisfactory situation that will eventually result in failure.

Machine deterioration frequently results in an increase in vibration and, if allowed to proceed, this will cause white metal bearings to fail by fatigue. One of the commonest types of journal bearing failure in high-speed machines is by fatigue, caused by out-of-balance loads or oil-film instabilities (see Fig. 5.7). Determining the level of vibration that is likely to cause bearing failure depends not only on the class of machine, but more particularly on the specific machine and its installation, and is largely a matter of experience. As a rough guide the double-amplitude of vibration in mm measured on the bearing housing should not exceed $100/N$ for new machines and $125/N$ for machines in service, where N is the frequency of rotation in rev/min.

More complete guidance is available in a number of sources. See, for example, VDI 2056: *Criteria for assessing mechanical vibrations of machines* (English translation Peter Peregrinus Ltd. 1971), BS 4675: 1976, *Mechanical vibration in rotating and reciprocating machines.*

VDI 2056 suggests limits for four groups of rotating machines:

Group K – small machines, e.g., electric motors up to 15 kW.

Group M – medium-sized machines, e.g., electric motors of 15–75 kW without special foundations.

Group G – relatively large machines on high-tuned, rigid or heavy foundations.

Group T – large machines on low-tuned foundations, e.g., turbomachinery assemblies.

Figure 7.5 shows the suggested vibration limits given by VDI for Group T machines. While plots like this are useful as an aid in assessing machine quality, the measurement of the vibration levels in new installations and the trend in service is the only reliable guide for particular machines. In addition to providing an indication of machine condition, analysis of vibration signals can also give clues to the source of the vibration in a particular problem.

In rolling bearings on the other hand, vibration is generated by deterioration of the bearing itself, rather than of the machine. While simple sounding sticks can provide a useful guide in skilled hands, more sophisticated monitors using ultrasound or vibration are available for bearing assessment, though even with these considerable experience is required in interpretation. With the aid of these methods it should be possible to detect the incipient failure of rolling bearings weeks or even months before complete collapse.

Increase in freedom of movement at thrust contacts can be shown by axial position indicators. These can be sufficiently sensitive to allow a machine to be tripped before there is any damage by the rotating element fouling the casing.

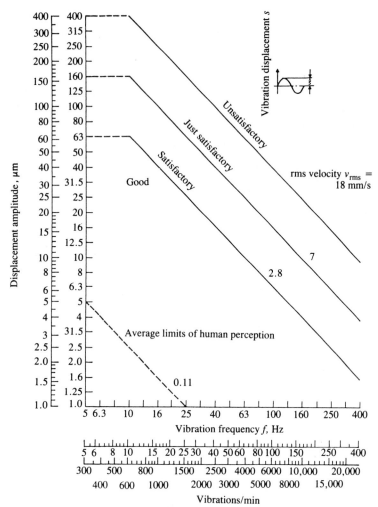

Fig. 7.5 Example of assessment limits of vibration behaviour for machine group T. Machines and turbomachines are mounted low-tuned on light foundations. (For non-harmonic vibrations, valid only for equivalent displacement amplitude.) (Source: VDI Guideline 2056 (Fig. 7). Printed with permission of the Verein Deutscher Ingenieure, Düsseldorf, Germany.)

Interpretation of the information from monitoring devices is a major problem in differentiating between genuine incipient failure and spurious effects. Where large critical machines are involved two completely independent methods should be used to remove any element of doubt. Table 7.1 gives possible techniques from which two methods can be selected for each of the parts to be monitored.

Another problem with monitoring and protective devices is that they may never be called on to perform during the life of the machine. This means they tend to get neglected and must, therefore, be reliable for long periods without attention. It is also important that they can be tested without the need to shut down the machine.

7.3 GEARS

7.3.1 Lubrication

At low speeds gears are most effectively lubricated by splash from a bath, but with increasing speeds this leads to excessive churning losses and it is thus necessary to spray oil on to the teeth and operate with a dry sump. Again at high speeds the oil tends to be flung off the teeth and is not available for lubrication at the point of mesh. As a guide, bath lubrication can be used up to pitch line velocities of about 12.5 m/s for spur, helical and bevel gears, for worm gears up to about 10 m/s.

At high surface speeds the oil spray is required as much to cool the gear teeth as to lubricate them. In fact the most effective way, as has already been mentioned, is to spray the teeth as they come out of mesh; this ensures the maximum cooling and means that no unnecessary oil is carried into the mesh where it would only increase the generation of heat by viscous shear losses.

There is no simple theoretical way in which the viscosity grade required for gear lubrication can be determined, though a number of more or less sophisticated empirical ones are available. In general, the viscosity recommended is inversely proportional to the pitch line velocity, as would be expected from the mechanics of fluid film lubrication, with some tolerance depending on tooth loading. The range of viscosity grades between the dashed lines in Fig. 7.6 gives viscosity grade values that are suitable for spur and helical gears operating in normal ambient temperatures. Straight

Table 7.1   Suggested monitoring scheme for large, critical machines

Component	Temperature	Monitoring technique		
		Vibration	*Detection of wear particles*	*Static movement*
Journal bearing	Thermocouple in contact with back of white metal in load zone	Housing velocity/ journal displacement		
Thrust bearing	Ditto			Shaft position indicator
Rolling bearing	Thermocouple in contact with stationary race	Velocity/acceleration	Magnetic plug	
Gears		Housing velocity	Magnetic plug	

mineral oils are suitable for most industrial gears, though where tooth loadings are high, oils with ep additives may be required (See Table 3.3).

7.3.2   Surface finish

With the low lubricant film thickness in gears, tooth finish is of great importance, particularly in running-in. In the addendum and dedendum of the tooth, relative sliding takes place during contact and high spots are removed by the running-in process already described. At the pitch line there is only pure rolling and this type of running-in does not occur; instead the high spots tend to carry the load and, with repeated stressing at successive tooth contacts, fatigue occurs and small pits are created. As the high spots are removed in this way the load is distributed more uniformly over the tooth so that the fatigue limit at the surface is no longer exceeded and pit formation ceases. With further running and gradual wear the existing pits may even begin to disappear. With hardened teeth that have a high fatigue strength this process may be very protracted and running-in may take up to two years. Pitch-line pitting is a natural process in gear teeth and should be no cause for alarm. However, if the gear is loaded beyond the design surface stress for the teeth (e.g., by misalignment), pitting occurs in a similar way except that the stable equilibrium condition is never reached and pitting continues until the tooth fractures. Unfortunately, it is not easy to distinguish between the initial pitting associated with run-in and the early stages of destructive pitting. In order to separate the two cases it is necessary to take replicas of specific teeth at intervals so the progression can be determined. Suitable replicas can be made by using a cast of silicone rubber or by coating the tooth with engineer's blue and taking a print with a strip of Sellotape.

An illustrated booklet showing pitting and other forms of gear tooth damage has been published by the American Gear Manufacturers' Association (*Nomenclature of gear-tooth wear and failure*, AGMA, Washington).

7.4   CYLINDERS OF RECIPROCATING COMPRESSORS

The lubrication of the cylinders of reciprocating compressors presents a number of interesting features. Compressors with trunk pistons rely on

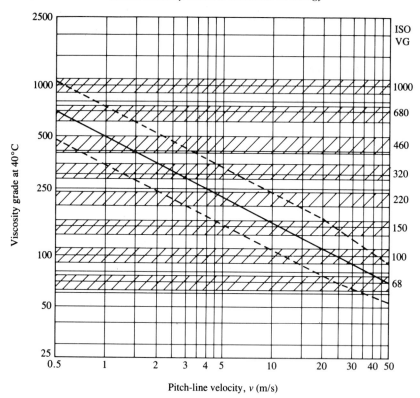

Fig. 7.6   Viscosity grade requirement for spur and helical gears as a function of pitch-line velocity. The ISO viscosity grades are superimposed. (Source: Based on Fig. 28, *The Lubrication of Industrial Gears*, 1967 (*Shell International Petroleum*))

splash from the crankcase for cylinder lubrication, but with crosshead machines the cylinders require an independent supply of lubricant. Both the viscosity grade and the rate of feed have to be determined. The situation is unusual in that intimate contact occurs between the lubricant and the gas being handled under conditions of relatively high temperature and pressure, so that consideration has to be given to possible interactions between the gas and the lubricant that can result in either physical or chemical changes in the latter.

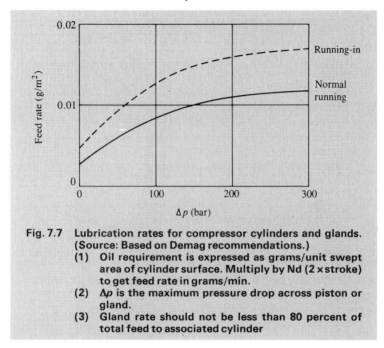

**Fig. 7.7   Lubrication rates for compressor cylinders and glands.
(Source: Based on Demag recommendations.)**

   **(1)   Oil requirement is expressed as grams/unit swept
          area of cylinder surface. Multiply by Nd (2 × stroke)
          to get feed rate in grams/min.**

   **(2) Δp is the maximum pressure drop across piston or
          gland.**

   **(3)   Gland rate should not be less than 80 percent of
          total feed to associated cylinder**

The aim, as in all lubrication situations, is to provide a fluid film between piston rings and cylinder. This is developed hydrodynamically by the movement of the piston and maintained at the ends of the stroke by squeeze-film action. A certain proportion of the oil is carried out of the cylinder with the discharged gas and the feed rate should be just sufficient to maintain an adequate film of oil on the cylinder walls. It is not possible to calculate the feed rate theoretically and this has to be found empirically by starting off with an excess and then, if it is important to use the minimum to avoid contamination of the gas, gradually reducing the rate until, on inspection, no oil is collecting in the valve pockets, while the cylinder walls are still sufficiently wet to stain a piece of cigarette paper. Figure 7.7 shows the recommendations of one compressor manufacturer given in terms of the feed rate in grams of oil per square metre of swept cylinder wall surface as a function of pressure. This gives a safe quantity and pro-

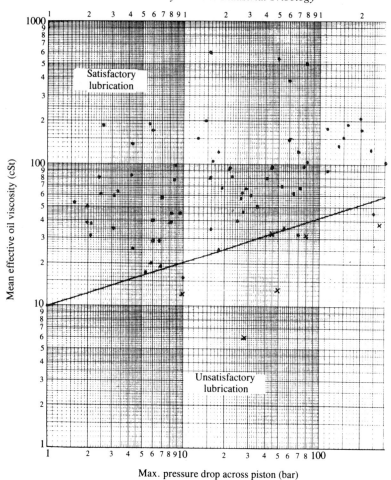

Max. pressure drop across piston (bar)

Fig. 7.8   Calculated effective oil viscosities as a function of pressure drop
  ● satisfactory lubrication
  ✕ unsatisfactory lubrication

vides a good starting point if it is desired to use the minimum possible to reduce contamination of gas. It is a good policy to use double the rate given in Fig. 7.7 during the run-in period as this helps to reduce the heat generated and wash away the running-in wear products.

Little information is available on the oil viscosity requirements for cylinder lubrication. Figure 7.8 gives guidance on the minimum effective oil viscosity required for satisfactory lubrication as a function of the maximum pressure drop across the piston, the effective temperature being taken as the mean between suction and discharge. When compressing gases with a high solubility in mineral oil (e.g., hydrocarbon gases, ammonia, carbon dioxide) it is necessary to make some allowance for the effect this has on reducing the oil viscosity. An empirical method for this has been obtained from operating experience on a wide range of gas compressors (for further details see D. Summers-Smith, Selection of lubricants for reciprocating gas compressors. *Proc. Instn mech. Engrs*, 1968). Additional experience obtained since this paper was published has confirmed that this method provides a reliable guide to lubricant viscosity selection.

CHAPTER 8

Fire and Explosion Hazards with Mineral Lubricating Oils

Though not highly flammable, mineral hydrocarbon oils can cause fires and explosions:

- *crankcase explosions*
- *hydraulic system fires*
- *air compressor fires and explosions*
- *pneumatic system fires*

Understanding the mechanism is the key to the elimination of the hazard.

Although mineral lubricating oils are not highly flammable materials there are a number of circumstances on industrial plant when they can give rise to serious fires and violent explosions. All engineers associated with the design and operation of plant should be aware of these circumstances so that steps can be taken to reduce the risks to a minimum.

8.1 CRANKCASE EXPLOSIONS

Figure 8.1 shows the flammability region for a typical hydrocarbon lubricant; this shows a minimum spontaneous ignition temperature of approximately 270°C at 12 percent oil by weight. Also shown in Fig, 8.1 is a typical vapour saturation curve for a medium viscosity mineral oil. Oil vapour is present in the air space of diesel engine crankcases, reciprocating compressors or steam engines, and gearboxes, but under normal operating conditions the mixture is well away from the flammability region. Even if a hot spot occurs as a result of parts rubbing together and the oil splashing on this is vaporized, the equilibrium vapour concentration curve does not pass through the flammability region and no ignition should occur. Away

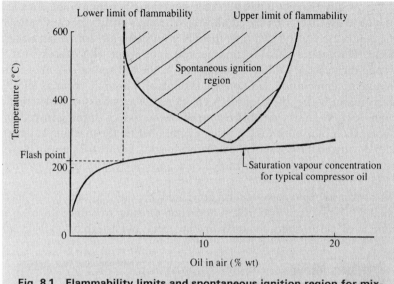

Fig. 8.1   Flammability limits and spontaneous ignition region for mixtures of lubricating oil vapour in air

from the heat, the oil condenses into a fine mist; it should be noted that air/oil mixtures have the same flammability characteristics as oil vapour/air mixtures. By the time that the temperature at the hot spot has risen above the minimum ignition point the mixture in the air space is too rich to ignite. However, if there is starvation of oil to the hot spot or a crankcase door is opened, admitting air and diluting the mixture, an explosion can occur with flame emission.

The primary explosion creates a pressure depression in the crankcase and, when this has occurred on the oil-rich side of the flammability region, a more violent secondary explosion can occur in through-the-turbulent mixing conditions created by the air flowing back into the crankcase.

WARNING

If blue 'oil smoke' is seen coming out of a crankcase or gearbox, the machine should be stopped and allowed to cool down before opening.

Frequently the first indication that something is wrong is the emission of blue 'oil smoke' from the crankcase. When this occurs the conditions in the crankcase are too rich (i.e., there is insufficient air) for an ignition to occur. The danger is that if an inspection door is opened to check what is wrong the necessary dilution occurs and, if the hot spot is still above the minimum temperature, ignition will take place. This may occur as long as thirty minutes after the door has been opened. Anyone associated with the running of machines should be warned of the risk of opening inspection doors when blue oil smoke is being emitted before the machine has been shut down and the hot spot allowed time to cool. Figure 8.2 shows a blown out cast iron crankcase door caused by a crankcase explosion in a reciprocating compressor.

Crankcase explosions can be prevented by the addition of an inert blanketing gas, such as nitrogen or carbon dioxide, to reduce the oxygen content of the air space to less than 10 percent, but as explosion incidents

Fig. 8.2   Inspection door on reciprocating compressor crankcase blown out as a result of crankcase explosion

are rare it is difficult to ensure that such a condition is maintained. The maximum pressure that is created by an explosion at the stoichiometric composition is about eight bar and, if the crankcase is sufficiently robust to withstand such a pressure, no action need be taken provided any vent covers are suitably trapped. Where this is not the case, the best protection is obtained by fitting a crankcase relief valve with a flame arrester, e.g., a fine wire mesh gauze that is kept wet by oil splash. The recommended relief valve area is $12 \, mm^2/1000 \, mm^3$ free space in the crankcase.

PROTECTION AGAINST CRANKCASE EXPLOSIONS

Fit crankcase relief valves:
relief area – $12 \, mm^2/1000 \, mm^3$ free air space in machine

It should be noted that although such incidents are usually referred to as crankcase explosions they can also occur in other systems, such as gearboxes, where there is an enclosed space of air above the oil.

8.2   HYDRAULIC SYSTEM FIRES

Large jets of cold oil sprayed on to a hot surface are unlikely to ignite; either the oil has sufficient heat capacity to reduce the temperature below the ignition limit, or the air/oil vapour mixture created at the hot spot is too rich for ignition. The feature of a hydraulic system is that the hydraulic medium is at high pressure. Most industrial hydraulic systems use mineral hydraulic oils, and if a leak occurs a finely atomized spray of oil is emitted. Such a spray impinging on a hot surface does not have the heat capacity to cool the ignition source, nor is there sufficient oil to produce an equilibrium over-rich mixture so that the conditions for ignition occur.

HYDRAULIC SYSTEM FIRES

The risk with a high-pressure hydraulic system is that a leak gives rise to a finely atomized spray that readily ignites if it impinges on a high temperature source.

Where high-pressure hydraulic systems occur adjacent to a source of ignition the only way of breaking the fuel–oxygen–ignition triangle is to use a non-flammable hydraulic medium. No satisfactory completely non-flammable medium is available for high-pressure systems and the option is to use a so-called fire-resistant fluid. Two types are available: aqueous based, e.g., oil–water invert emulsions containing about 40 percent water, polyglycol–water mixtures, and 'synthetic' fluids such as phosphate esters. The former, which depend on the water evaporating to form a steam blanket, are limited in use to about 60°C to avoid excessive loss of water and require regular checking to ensure that the water content is maintained. Suitably inhibited phosphate esters can be used up to about 140°C. Both classes of fire-resistant fluids tend to have somewhat inferior lubricating properties to mineral oils and account has to be taken of this when designing or converting a system; in addition, they may place special demands on seals and paints.

Casting shops and coal mines are obvious places where hydraulic system fire risks exist. In most industrial plant mineral hydraulic oils present no hazard.

8.3   LAGGING FIRES

Drips of oil on to hot lagged pipes may result in ignition even if the temperature is well below the minimum ignition limit given in Fig. 8.1. Porous lagging material has a large surface area so that any oil leaking on to it spreads out to form a very thin film. Such a thin oil film in contact with air begins to oxidize and, as the oxidation is an exothermic reaction, heat is evolved. With lagging materials of low thermal conductivity, the temperature can rise to the spontaneous ignition region even if the lagging itself is below the critical temperature. In the confined air space in the lagging, over-rich conditions are probably created, but smouldering may occur at the surface. The main hazard at this stage occurs with stripping away the lagging, giving increased access of air and probably a significant chance of fire as more hot oil is exposed. The minimum lagging temperature required to give a risk of ignition is not accurately defined but, depending on the precise conditions, is probably in the range 150–200°C.

LAGGING FIRES

Mineral lubricating oil dripping on to porous lagging can ignite spontaneously if the lagging temperature is at least 150–200°C.

It should be noted that the water-based, fire-resistant hydraulic fluids are not immune to lagging fires, the water merely evaporating off leaving behind a flammable residue.

Where possible, lagging should be protected by metal sheet cladding or by an impermeable finish. Protection against lagging fires is then a matter of good housekeeping, i.e., promptly repairing leaking oil joints and then replacing cladding after maintenance work. In areas where there are liable to be oil leaks, crimped metal foil, though less effective as an insulating material, gives no risk of fire as it lacks the wicking action of porous lagging. Alternatively, it may be better to avoid the use of lagging in places where oil drips regularly occur as this will ensure the equilibrium condition of Fig. 8.1 and the circumstances for ignition do not arise.

8.4 AIR COMPRESSOR FIRES AND EXPLOSIONS

So far, cases at normal atmospheric pressure have been considered where ignition of oil vapour cannot occur under equilibrium conditions. This is not the case, however, at elevated pressure. Figure 8.3 shows the spontaneous ignition region for a 12 percent by weight oil vapour/air mixture as a function of pressure. It will be seen that at a pressure of about 1.3 bar the vapour pressure curve enters the spontaneous ignition region and that at 100 bar the ignition temperature falls to about 190°C.

AIR COMPRESSOR FIRES

Mixture of mineral vapour or mist and air ignites spontaneously at about 240°C at 1.3 bar falling to 190°C at 100 bar.

Prevention of ignition depends on limiting the oil concentration.

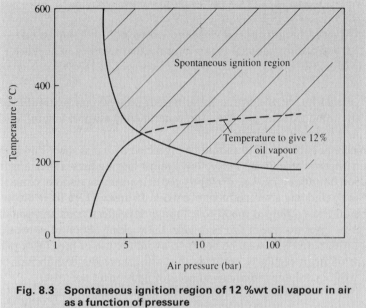

Fig. 8.3 Spontaneous ignition region of 12 %wt oil vapour in air as a function of pressure

In normal circumstances the conditions for ignition do not exist in air compressors, even if the ignition temperature is reached, as the amount of oil is too low to form a flammable mixture. Fires do occur, however, both in oil-flooded rotary and reciprocating air compressors; it is necessary to understand why this happens, if fires, and the even more damaging explosions that can follow air compressor fires, are to be avoided.

In an oil-flooded rotary air compressor, oil is passed through the compressor to absorb the heat of compression and allow high compression ratios. In the event of a mechanical rub in the compressor sufficient oil should be present to prevent ignition.

Excess oil is removed from the delivery oil by a separator, the air stream impinging on an inclined plate to remove the coarse oil droplets and then through a glass wool filter to remove the fine mist. A flammable mixture can be present in the separator, but discharge temperatures are limited to 100–120°C, so the conditions for ignition should not exist. Downstream of the separator the oil content is reduced to less than 50 ppm, well below the flammability region and below the saturation vapour pressure so that it does not condense out on the pipe wall. Fires can only occur as the result of a fault condition.

Interruption or reduction of the cooling oil through the compressor not only allows the temperature in the separator to rise, but could eventually allow the formation of a flammable mixture in the compressor that could be ignited by a mechanical rub. The oil in the oil mist filter is present as a thin film and, just as the case with lagging fires, will oxidize at an increasing rate when the discharge air temperature rises. If this is allowed to continue the heat of oxidation can raise the temperature to the spontaneous ignition point and a fire will result. Oil-filled rotary air compressors are fitted with high temperature trips designed to protect against such fires, but the rate of temperature rise can be such that the rate of response of such trips may not be rapid enough to shut the machine down before a fire occurs. Regular checking to ensure that the oil quantity in the system is kept at the proper level is a much better protection against fires. The reason for any abnormal topping-up requirement should be thoroughly investigated.

Rapidly moving droplets can carry static charges and it is important that the filter basket is earthed to the casing of the separator so that there is no build up of static voltage. Fires have occurred as the result of sparks between the filter basket and the casing igniting the oil mist/air mixture in the separator when the earthing strap has been omitted. Figure 8.4 shows a hole in the discharge piping of an oil-filled rotary compressor following a fire in the separator ignited by a spark between the filter casing and the body of the separator.

Fires in oil-lubricated reciprocating air compressors arise from a completely different mechanism. The rate of oil feed to the cylinders is well below the flammability limit. However, this oil is subject to severe oxidizing conditions at the discharge conditions of temperature and pressure. Initially the oil increases in viscosity with oxidation and, if oxidation is allowed to continue, it eventually is transformed to a solid carbonaceous mass. Such carbonaceous deposits are porous and tend to absorb oil that is then subjected to oxidation. Initially any heat generated in the oxidation process is partly conducted through the pipework and partly carried away by the discharge air. However, as the deposit thickness builds up heat loss to the pipe falls and the temperature rises until spontaneous ignition occurs. This gives rise to a serious fire as the oil-soaked carbonaceous

Failed back pressure/check valve (brass)

Air flow ⟵

Fig. 8.4 Holed delivery line from oil-filled rotary air compressor as a result of fire

deposit burns in the hot stream of air; this can give rise to temperatures of about 1000°C, weakening the pipework so that it bursts under the discharge pressure and burning out aftercoolers (Fig. 8.5). It is important that the oil carried from the compressor in the delivery air is swept through the hot discharge pipework before deposit formation can take place. This can be achieved by a combination of good design, with the pipework arranged so that gravity assists the flow of the oil through the system, choosing an oil specially designed to resist deposit formation, and limiting discharge temperatures to prevent excessive rates of oxidation. It is also important not to use unnecessarily high rates of lubrication in the cylinders to reduce the risk of build up of oil in the discharge system. The oil feed in crosshead compressors is controlled by a separate lubricator, and overlubrication is easily avoided. With trunk piston machines, where cylinder lubrication is provided by splash from the crankcase controlled by the piston rings, any abnormal rate of crankcase top up should be investigated to ensure that it is not the result of carryover into the discharge system.

Fig. 8.5   Burnt-out aftercooler following fire in oil-lubricated reciprocating compressor

A secondary effect of air compressor fires is that they can initiate explosions in oil-wetted pipework. This occurs when the fire generates a shock wave that is transmitted down the pipework. The shock wave strips oil from the pipe and explosions occur when the conditions at the shock wave front enter the spontaneous ignition region. These pipeline explosions, which occur under the good mixing conditions of turbulence in the pipe, are detonative in character, i.e., the explosion pressure wave front exceeds the velocity of sound, and are extremely violent, causing brittle fracture of the pipe with the projected fragments presenting a serious hazard. Some impression is given of the seriousness of such occurrences by an incident that took place in Germany in 1963, which led to nineteen people losing their lives. Not all air compressor fires result in pipeline explosions, but the risk is high and is the major reason why steps should be taken to prevent any occurrence of air compressor fires.

WARNING

Lubricated reciprocating and oil-filled rotary air compressors should not be coupled to the same compressed air system.

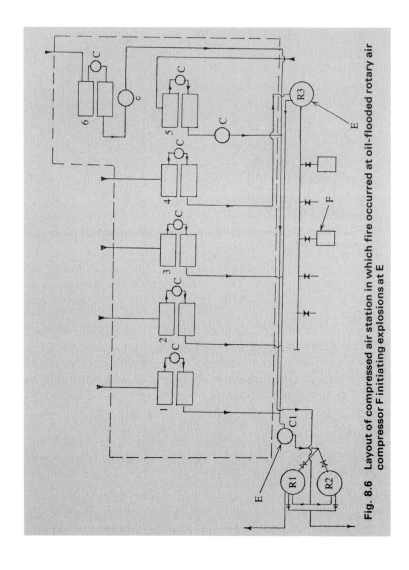

Fig. 8.6   Layout of compressed air station in which fire occurred at oil-flooded rotary air compressor F initiating explosions at E

Pipeline explosions should not occur with oil-filled rotary air compressors, as insufficient oil is carried forward to wet the pipes. However, as fires are more probable in rotary compressors, they should not be connected to the oil-wetted delivery systems of oil-lubricated reciprocating compressors. Figure 8.6 gives the layout of a compressed air system with six reciprocating compressors where oil-filled rotary compressors were used to cover peak periods of high demand. A fire that occurred in a rotary machine initiated two secondary explosions in the pipework from the reciprocating compressors, causing serious damage and outage of the whole system.

8.5   EXPLOSIONS IN HIGH-PRESSURE PNEUMATIC SYSTEMS

Explosions can occur in high-pressure pneumatic systems containing small amounts of oil in a similar way to that described in oil-wetted pipes, when valves are opened quickly. Either the high pressure air acts like a piston and compresses the air on the low pressure side adiabatically, raising the temperature to the ignition limit, or a shock wave is created and an explosion occurs by a similar mechanism to that described for compressed air systems.

Incidents of this type have occurred in testing high-pressure valves, when the sudden admission of air at 450 bar has ignited the petroleum-based preservative in the valve. Similarly, explosions have occurred on opening valves in high-pressure pneumatic systems, where the compressed air was provided by an oil-lubricated reciprocating compressor that had carried oil forward into the pneumatic system and at a dead-end on a combined hydraulic/pneumatic system used for the positioning of the drill on a drilling platform.

It has to be stressed that mineral lubricating oils are not highly flammable materials and do not present a fire or explosion risk in normal operation. None of the incidents described above are common, but they do occur and it is important that engineers are aware of the circumstances that give rise to them in order that incidents that can cause serious injury to personnel or damage to plant are avoided.

CHAPTER 9
Dynamic Seals

Satisfactory lubrication is basic to the successful performance of dynamic seals on rotating and reciprocating
shafts:

soft-packed glands
automatic packings
mechanical packings
mechanical seals
piston rings

Lubrication is mostly in the boundary or mixed regimes with
minimization of heat generation and attention to cooling the
key points.

The subject of dynamic sealing is closely related to that of bearings.
Again we are concerned with surfaces moving relatively to each other,
either in contact or with a fluid between them. In the case of bearings the
aim is that there should be no damage to the surface and that the friction
loss should be minimized. In sealing, the objective is that the leakage of
the contained fluid at the point where movement occurs should be kept
to a minimum. The major difference is that whereas in bearings the lubricant is chosen to give the optimum conditions, using a large flow rate
through the bearing to conduct away the frictional heat and prevent
excessive temperature rise, in most seals the 'lubricant' is the sealed fluid
and, apart from the 'clearance seal' (e.g., throttle bushes, labyrinths) that
has no direct bearing equivalent, there is little flow through the seal to
remove the heat.

As with bearings, there are basically two types of dynamic seal: fluid
film and rubbing. The choice of type and the detailed design is influenced
by the type of fluid to be sealed – gas or liquid – and the nature of the
relative motion – rotary or reciprocating.

FACTORS IN SEAL TYPE SELECTION		
Sealed fluid	*Gas*	*Liquid*
Rotary shafts	Mechanical seal Throttle bush or labyrinth Floating bush	Lubricating 'O' ring Mechanical seal Non-lubricating Mechanical seal
Reciprocating shafts	Piston ring Mechanical packing	Automatic packing Soft packing Soft-packing

9.1 SEALING OF LIQUIDS

Contact seals are used almost exclusively for the sealing of liquids, some-times in conjunction with a throttle bush to reduce the pressure. In the majority of cases the sealed liquid is used to lubricate the contact. Soft-packings and automatic packings are used for reciprocating applications, 'O' rings, and mechanical seals for rotary applications.

9.1.1 Reciprocating seals: soft-packed glands
Soft-packed glands find their principal application on the plungers of reciprocating pumps. The principle is that a resilient packing material is loaded into an annular space round the shaft and compressed by tightening up a gland so that the packing is pressed radially against both the shaft and the wall of the stuffing-box (Fig. 9.1). The prin-ciple is simple, but satisfactory performance depends on attention to detail.

A wide variety of materials is available for soft-packings, ranging from mineral, vegetable, and synthetic fibres that are braided or plaited into a square-section rope (Fig. 9.2) to compacts of crimped metal foil round a fibre core. The materials are usually impregnated with a lubricant – vegetable or mineral oil, grease or solids such as graphite and ptfe. This gives rise to a very large number of products, both in the variety of materials of construction, cross-section, and the density of the final pack-ing, but there is actually little genuine variety as far as performance is con-cerned. Table 9.1 gives guidance on the field of application of the principal soft-packing materials.

CONSIDERATIONS IN SOFT-PACKING MATERIAL SELECTION

- Chemical resistance to the fluid being sealed.
- Thermal resistance to the operating temperature.
- In-built lubricating properties to cover the run-in period.

The first two above considerations are self-evident; the last may require some explanation. Soft-packings operate in the boundary lubrication regime: in normal running, lubrication comes from the sealed liquid, or when leakage of this has to be prevented, by an acceptable liquid fed to the centre of the packing via a lantern ring (Fig. 9.1). It takes some time for the packing to settle down and for correct lubrication conditions to be established. The function of the lubricant in the packing is to cover this bedding-in period. Once this lubricant has been lost from the surface, there is no further supply from the bulk of the packing and separate lubricant has to be applied to the rubbing contact.

Resilience in the packing is required so that it deforms to fill the annular space in the stuffing-box, but the resilience is limited and for satisfactory performance the packing must fill the annular space during fitting as closely as possible. This means that rings should either be die-pressed to

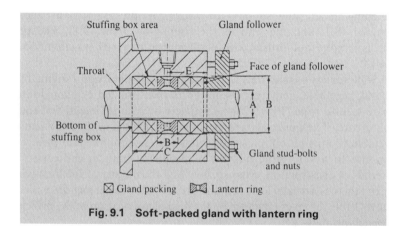

Fig. 9.1   Soft-packed gland with lantern ring

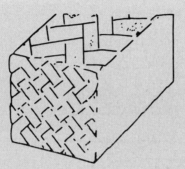

Fig. 9.2   Braided construction helps to give resilience

Table 9.1   Selection of soft-packing material

Type of packing	Applications

Cotton – grease-impregnated	Water and aqueous-based liquids
	    Max. temperature 80°C
	    Max. sliding speed 7.5 m/s

Hemp – grease-impregnated with graphite	Hydrocarbons
	    Max. temperature 80°C
	    Max. sliding speed 7.5 m/s

Crimped metal – aluminium, lead or white metal foil on mineral fibre core	Hot hydrocarbons
	    Max. temperature:
	        200°C with lead or white metal
	        500°C with aluminium

Aromid fibre	Strong acids and alkalis (pH 2–12)
	    Max. temperature 250°C
	    Max. sliding speed 15 m/s

Exfoliated graphite	Strong oxidizing acids (e.g., nitric acid)
	    Max. temperature 500°C
	    Max. sliding speed 10–15 m/s

Notes
1   Asbestos impregnated with ptfe or graphite gives excellent results with steam and strong acids, but is now proscribed on health grounds.
2   Co-plaited ptfe with natural or synthetic fibres is not a preferred material for high-pressure applications as the ptfe flows under pressure and the packing loses its resilience.

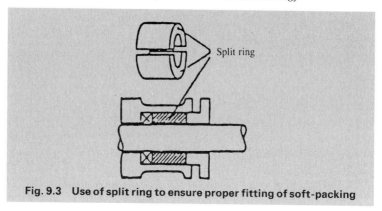

Split ring

Fig. 9.3   Use of split ring to ensure proper fitting of soft-packing

size or individually pressed into place, preferably by using a split collar that can be tightened up with the gland follower (Fig. 9.3). To assist in applying uniform pressure a three-bolt gland plate is preferred.

It is common experience that sealing is effected by the one or two rings on the atmospheric side. Recognizing this, modern practice is to limit the packing to 2–4 rings, giving adequate sealing performance without the excessive frictional heating that occurs when a larger number of rings is used. Again, because of the problem of removing heat from the contact, it is found that rings of limited cross-section, about 5 mm, give the best results.

KEY FACTORS IN SOFT-PACKED GLAND DESIGN

Number of rings of packing   –   4–5
Packing cross-section   –   5 mm
Shaft surface finish   –   0.2–0.4 μm
Packing technique   –   fit each ring individually

Effective lubrication is the key to the success of soft-packed glands. This depends on good fitting, controlling the leakage to an acceptable level (normally about 1–2 drops/minute), but avoiding overtightening that results in the failure of lubrication, overheating, and scoring of the shaft. Regular attention is required to control the leakage at the correct level by tightening the gland as the packing wears. Exfoliated graphite packing is somewhat different from the other types of soft-packings in that it relies on elasticity rather than resilience to give a close fit in the stuffing-box. Being a self-lubricating material with a higher thermal conductivity than conventional soft-packing materials, it can operate with zero or at least very low leakage without excessive friction and overheating.

Best results are obtained with soft-packed glands when the run-out is less than 0.05 mm total indicator reading. The function of the packing should be limited to sealing with guidance of the rod effected by the crosshead and neck bush.

9.1.2 Reciprocating seals: automatic packings

The automatic packing consists of a series of rubber or plastic 'U' rings or Chevron-rings that are energized by the pressure of the sealed fluid and forced against the plunger and stuffing box wall; these require good lubrication. In operation a comparatively thick film of the sealed liquid (ca. 0.25 mm) is drawn out on the suction stroke and, provided this film remains intact on the rod and does not retract into discrete droplets, much of it returns on re-entry into the pressure side maintaining good lubrication conditions. Leakage, i.e., the difference in film thickness between the withdrawal and re-entry strokes, tends to be quite low.

Because of the restriction on wetting the rod, the use of automatic packings is limited to sealing liquids of low surface tension. Rubber automatic packings have been used successfully up to at least 250 bar on hydrocarbons that have adequate viscosity to provide lubrication and are not so volatile that they evaporate from the rod on withdrawal. Automatic packings are not suitable for aqueous solutions.

9.1.3 Rotary seals: 'O' rings

By far the commonest type of seal, though the least conspicuous, is the rubber 'O' ring. For satisfactory performance it depends on the generation of a lubricating film (almost certainly an elasto-hydrodynamic lubricating film). Its use on industrial machines is almost entirely limited to oil seals on gearboxes and similar applications, where it does an excellent, and for this reason unnoticed, job.

9.1.4 Rotary seals: mechanical seals

Soft-packed glands have almost entirely been replaced by mechanical seals for the sealing of rotating shafts. In the mechanical (or radial face) seal the sealing contact is between two radial faces, one fixed to the shaft or vessel, the other rotating with the shaft. One or both of the faces is mounted flexibly; one face is free to move axially and is spring-loaded to ensure that the two faces are pressed together at all times (Fig. 9.4). The faces may be in solid contact, but the best results are obtained when there is a fluid lubricating film between them giving some separation, though for effective sealing this has to be very thin (0.25–2 μm).

The successful operation of mechanical seals depends on the maintenance of stable conditions at the sealing interface. Investigations of seal performance by the statistical technique of Weibull analysis have shown that most seal failures occur by infantile mortality (early-life failures) or follow a random pattern (mid-life failures). Early-life failures are a consequence of inadequate design (choice of the wrong seal, seal materials, or sealing arrangement) or poor fitting. Mid-life failures are the result of external factors, e.g., loss of the interface film by vaporization, vibration, solids in the sealed liquid, hang-up of the floating face. Seals seldom, if ever, reach a predictable wear-out life. The reliable application of mechanical seals depends on a recognition of these features, particularly the extreme thinness of the interface lubricating film.

Experimental results suggest a typical Stribeck-type curve for mechanical seal operation, though the reproducibility is poor, largely because of thermal and mechanical distortion effects on the geometry of the contact (Fig. 9.5). This type of plot, shown schematically in Fig. 9.6, is, however, useful in visualizing the operating conditions at the face contact. In practice it is found that most seals on reciprocating pumps operate about the minimum of the curve, at duty parameter values of 10^{-8} to 10^{-7}, i.e., in the transition region between fluid film and boundary lubrication.

DUTY PARAMETER FOR MECHANICAL SEALS

$$G = \eta v / p'$$

More practical considerations suggest that there are two operating limits to mechanical seals: a PV (pressure × velocity) limit corresponding

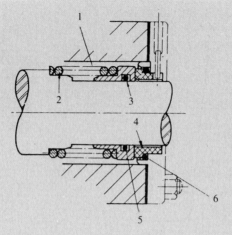

(a) Pusher-type mechanical seal with rubber
'O'-ring secondary seal

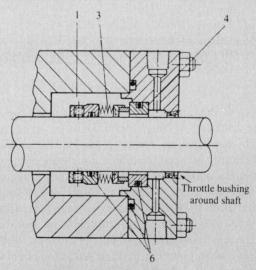

(b) Metal bellows-type mechanical seal

Fig. 9.4 Basic features of a mechanical seal

1	Seal chamber	**5**	Stationary seal
2	Spring	**6**	Stationary secondary
3	Dynamic secondary seal		seal
4	Rotating seal face	**7**	Bellows

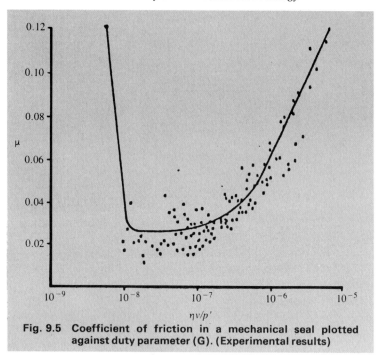

Fig. 9.5 Coefficient of friction in a mechanical seal plotted against duty parameter (G). (Experimental results)

to the breakdown of the fluid film, and hence involving the dry running properties of the seal face pair, and a thermal limit based on the loss of the lubricating film by vaporization. The latter can be expressed in terms of a ΔT limit (the difference between the bulk temperature of the liquid in the seal chamber and the boiling point, or bubble point in the case of mixtures, of the liquid at the sealed pressure) to ensure against vaporization of the interface film. Both these limits can be obtained experimentally for particular designs of seals and combined together in an operating envelope (Fig. 9.7). Where it is essential to prevent any external leakage of the pumped fluid, e.g., when pumping toxic or hazardous liquids, a double mechanical seal can be used. In this arrangement two seals are mounted back-to-back in the seal chamber with an acceptable barrier liquid supplied between them at a pressure about 1 bar above that of the pumped

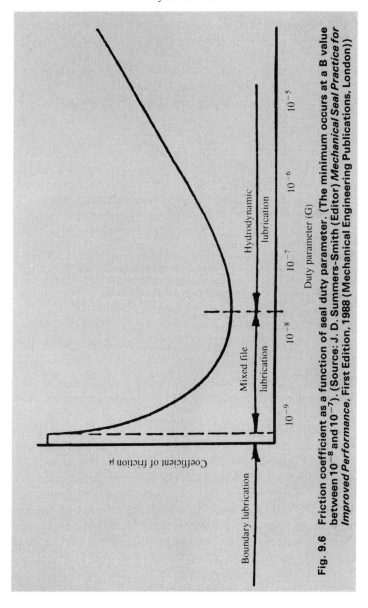

Fig. 9.6 Friction coefficient as a function of seal duty parameter. (The minimum occurs at a B value between 10^{-8} and 10^{-7}). (Source: J. D. Summers-Smith (Editor) *Mechanical Seal Practice for Improved Performance*, First Edition, 1988 (Mechanical Engineering Publications, London))

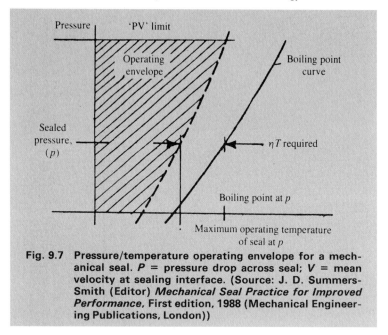

Fig. 9.7 Pressure/temperature operating envelope for a mechanical seal. *P* = pressure drop across seal; *V* = mean velocity at sealing interface. (Source: J. D. Summers-Smith (Editor) *Mechanical Seal Practice for Improved Performance*, First edition, 1988 (Mechanical Engineering Publications, London))

liquid in the seal chamber. Double mechanical seals can also be used with liquids containing solids in suspension to prevent penetration of the solids between the seal faces.

Seal performance depends not only on the seal design, but also on the design of the seal chamber; taken together this is best described as the 'seal arrangement'. Figure 9.8 (which is taken from *Mechanical Seal Practice for Improved Performance*, Mechanical Engineering Publications Limited, 1992) shows a selection chart that takes into account the main operating parameters. This provides useful first guidance in selecting the best seal arrangement. An overriding principle is that the simplest possible arrangement should be chosen: increasing complexity leads not only to increasing cost, but also to an increased risk of failure in service. The use of a dead-ended seal chamber, avoiding the need for the added complexity of external connections, is the best solution if it can be applied.

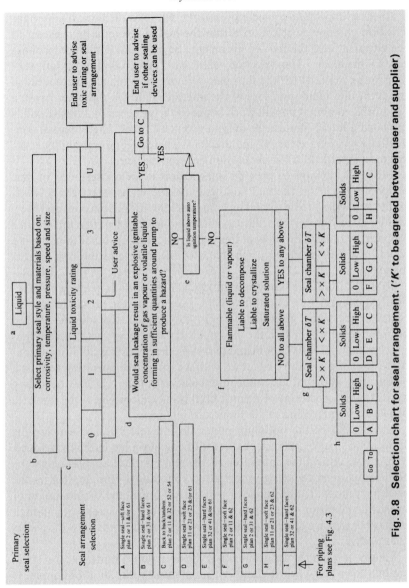

Fig. 9.8 Selection chart for seal arrangement. ('*K*' to be agreed between user and supplier)

Another important consideration is the selection of the seal materials. The first criterion in selection is that the material should be resistant to the chemical and thermal environment. Beyond this some compromise has to be made. The selection of a hard material and a soft self-lubricating material for the face pair gives the best chance of recovery in the event of a temporary loss of the interface film. On the other hand, a hard/hard combination will work better in the presence of abrasive solids provided a liquid film can be maintained at all times. High thermal conductivity will reduce the ΔT requirement; whereas good wettability is the key to successful fluid film generation between the faces.

Table 9.2 attempts to provide some rationale for the selection of face materials, based on thermal conductivity (the higher the better), wettability (the ability to be wetted by liquids of high surface tension) and thermal shock resistance, the ability to withstand intermittent film breakdown followed by quenching without thermal stress cracking; other things being equal, materials at the top right hand of the table give the best performance.

Table 9.2 is intended for qualitative guidance, but it is in line with practical experience. For example, where, because of the requirements for chemical resistance, glass-filled ptfe has to be used with alumina, it has been found necessary to limit the application to liquids with surface tension less than 0.2×10^{-3} N/m and to restrict the operating conditions compared to that that can be obtained with a carbon face operating against a metal. The apparent optimum combination silicon carbide operating against tungsten carbide or against itself is only acceptable if the presence of a liquid interface film can be assured at all times.

Having chosen the optimum materials and seal arrangement for the application, the best chance of successful operation lies in the recognition of just how critical the conditions at the seal interface are and hence the need to maintain the design lubrication conditions. This emphasizes the need for venting, maintenance of seal flush if fitted, and the limitation of swash and vibration. The last two factors will both reduce seal life, though at the present stage of understanding it is not possible to express the effects quantitatively.

Finally, it is worth mentioning that the optimum conditions for a stuffing-box for a soft-packed gland, viz an annular width of about 5 mm, is not the optimum for a seal where a more generous seal chamber allows the

Table 9.2 Seal face material selection based on tribological properties

Quality rating	Thermal conductivity (W/m K)				
	<1	1–20	21–40	41–60	>60
(1) Good thermal shock resistance; wetted by liquids with surface tension $> 5.0 \times 10^{-3}$ N/m		Carbon–graphite		Ni-resist	SiC
(2) Moderate thermal shock resistance: wetted by liquids with surface tension 0.5×10^{-3} N/m max		Duplex stainless steel	Al_2O_3		WC
(3) Poor thermal shock resistance; only wetted by liquids with surface tension 0.2×10^{-3} N/m max	Glass-filled ptfe	Stellite			

Table 9.3   Current status of mechanical seal technology

Sealed pressure (bar)	Comment
0–1	Doubtful applications
1–10	Basic designs give good life
10–40	Basic designs give reduced life or special designs necessary for enhanced life
40–100	Custom built design necessary

use of more robust seals and better fluid flow conditions round the seal. Interchangeable housings do not give the best results and seal chambers should be designed with the type of seal in mind.

Table 9.3 gives an overview of the current status of mechanical seal technology. There is no reason why seals on process centrifugal pumps, using the best state-of-the-art selection of seal arrangement and maintaining the intended design conditions in the seal chamber during operation, should not give reliable trouble-free service. Table 9.4 shows how, in fact, mechanical seal performance on process pumps has improved as a result of greater understanding. Reference should be made to *Mechanical Seal Practice for Improved Performance*, Mechanical Engineering Publications Limited, London, 1992, for more detailed information on seal design and operating practice.

Table 9.4   Field surveys of mechanical seal performance on process pumps

No. of pumps (investigation period)	Industry sector	MTBF (years)	Source
292	UK process	1.2	Flitney and Nau (BHRA 1977)
319 (3.5 years)	Petrochemicals (1 site)*	3.1	Summers-Smith (1981)
1000 (10 years)	Petrochemicals (1 site)*	5.8	von Bertele (1989)

* Note the petrochemical site is the same one, the second survey extending the investigation to ten years
MTBF = mean time between failures

9.2 SEALING OF GASES

9.2.1 Clearance seals

In a clearance seal, a clearance is maintained between the sealing surfaces independently of the operating conditions. As the surfaces are not in relative contact there should be no wear and the performance should not alter with time. However, because of the positive gap, sealing efficiency is lower than with a contact seal. Leakage is a matter of fluid dynamics and has little to do with tribology, though for the sake of completeness in any discussion on dynamic sealing, clearance seals have to be considered along with the other types.

9.2.2 Labyrinth seals

With gases, a labyrinth design can be used, making use of the gas expansion and turbulence created in the labyrinth pockets to reduce the rate of flow. The flow rate through a labyrinth can be calculated fairly precisely and, in theory, a labyrinth could be designed for a specific leakage rate. In practice, however, it is seldom realistic to design a labyrinth on a rotating shaft for a leakage rate of less than 0.5 percent of the machine throughput. This limits the use of labyrinths on atmospheric seals to machines handling air or steam where the cost of the fluid allows this rate of leakage to be acceptable. Labyrinths are, however, frequently used as internal seals between the stage of centrifugal gas compressors, where the high leakage rate represents a machine inefficiency, but not a loss of gas. Labyrinths are also used on the pistons of reciprocating gas compressors as one method of operating without cylinder lubrication. Leakage rates with reciprocating labyrinth seals can be as high as 15 percent of the gas rate, but again this is internal leakage and only reflects on the volumetric efficiency of the machine. Reciprocating labyrinth seals present a difficulty in guidance of the rod to ensure that no touching takes place. The major application of reciprocating gas labyrinths has been in oxygen compressors where it is essential to exclude any potentially flammable materials.

Mechanical seals, using hydrodynamic effects to produce a fine clearance between the two sealing faces, have been developed to operate directly as gas seals. Such seals tend to be rather specialized and expensive: the more common method of sealing gases where labyrinths are not acceptable is to use liquid barrier seals.

9.2.3 Liquid barrier seals

Where large leakage rates of gas cannot be tolerated, the normal practice is to use liquid barrier seals. These can be either a double mechanical seal or a multiple floating-bush seal, each design using an intermediate barrier liquid, small quantities of which leak into the gas being sealed. Double mechanical seals are mainly used for slow-speed applications, such as shafts of autoclave stirrers, and floating-bush seals on high-speed rotary compressors.

9.2.4 Double mechanical seals

Double mechanical seals mounted back-to-back with a barrier liquid injected between the seals can be used for sealing gases, the barrier liquid being supplied at a pressure about 0.5 bar above that of the sealed gas. As with all lubricated conjunctions, the aim in a mechanical seal is to generate a liquid film between the faces. As is shown in Fig. 9.6 this requires a duty parameter, G, between 10^{-8} and 10^{-7}. The use of a double seal in slow-speed applications gives the freedom to select a barrier liquid of sufficient viscosity to compensate for the slow speed and still obtain the necessary G value. On applications such as vertical autoclave stirrers, the barrier liquid can be conveniently circulated through the seal using a head tank with thermal siphon circulation, with the proviso of course that the barrier liquid must be compatible with the process.

9.2.5 Floating-bush seals

Floating-bush seals are the most common type of seal used on the shafts of rotary compressors. Figure 9.9 shows the general arrangement with the barrier liquid, normally the system lubricating oil, maintained at 5 m head above the pressure of the sealed gas by a head tank balanced to the seal chamber.

The use of floating bushes that offer no radial restraint on the shaft allows close clearances to be used and thus limits the internal leakage of the sealant into the process. Because of the low clearance the rings are lined with white metal, normally a tin-rich white metal, to avoid damage to the shaft in the event of rubs.

Two practical points should be noted in the application of floating-bush seals.

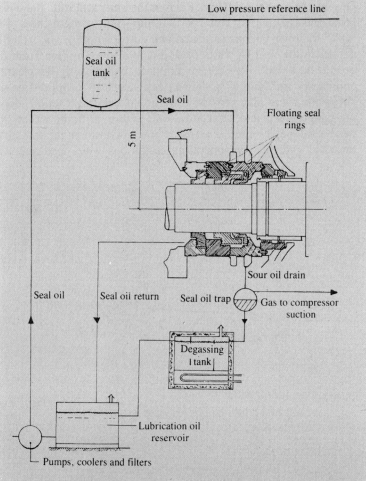

Fig. 9.9 Floating-bush shaft seal and seal oil circuit on centrifugal compressor

(1) Despite the pressure difference across the inner seal ring, there is still some leakage of gas. This has two important consequences. First, 'sour' hydrocarbon gases containing active sulphur can attack the copper in the tin-rich white metal lining of the ring (see Table 5.1), producing a deposit of copper sulphide that takes up the clearance and causes seal failure. In such cases a switch to a copper-free lead-rich white metal may provide a solution. Chemical reactions depend on temperature and concentration. Figure 9.10, in which shaft peripheral speed is used as an index of temperature in the seal, gives a practical guide to the operating conditions when lead-rich white

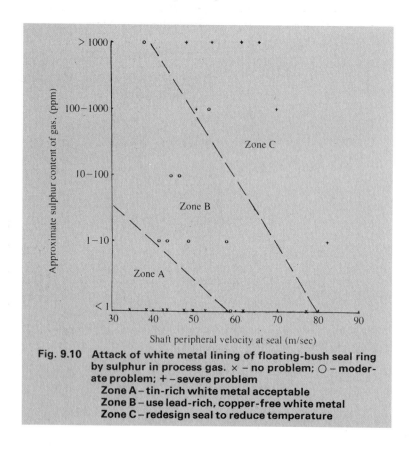

Fig. 9.10 Attack of white metal lining of floating-bush seal ring by sulphur in process gas. × – no problem; ○ – moderate problem; + – severe problem
Zone A – tin-rich white metal acceptable
Zone B – use lead-rich, copper-free white metal
Zone C – redesign seal to reduce temperature

metal seal linings should be used. The second consequence of leakage across the seal is that process gas will gradually build up in the gas space above the lubricant in the lubricant reservoir. In the case of hydrocarbon gas compressors, this can lead to the formation of a flammable mixture in the oil tank and a number of cases of explosions in the oil tanks of ethylene compressors have been reported, the ignition being caused by sparks through static build-up when the return oil is sprayed on to the surface of the oil. In such cases the oil should be returned through a stilling (dip) tube.

(2)   In most systems the oil leaking into the machines is recovered through let-down pots and returned to the systems after degassing, the gas being returned to the compressor suction. It has to be recognized that degassing is never 100 percent efficient and this means that there is a second mechanism causing some contamination of the lubrication system by process gas. This can be a problem with some gases, e.g., ammonia in ammonia synthesis gas compressors can cause lubrication problems by reacting with the oil additives. This type of contamination can be overcome by using vacuum degassing. Feeding a constant rate of oil through the degasser, making up the leakage with oil from the tank, has the added beneficial effect of completely drying out the oil.

9.2.6   Contact seals

Contact seals are used for sealing the pistons and piston rods in the case of double-acting compressors: these can either be lubricated or unlubricated using self-lubricating materials. Lubricated piston rings for industrial reciprocating compressors are made of grey cast iron or lead–bronze. Lead–bronze, white metal or grey cast iron (tinnized to assist run-in) are used for mechanical rod packings that consist of a number of pairs of opposed tangential and radial cut rings. Hardened rods are required with lead–bronze (250 HB min.) and cast iron (400 HB min.). Lubricant feed rates have already been discussed in Chapter 6. A hydrodynamic lubricating film is generated during the stroke, but this tends to break down to boundary lubrication at top and bottom dead centre where there is no relative motion. As stated in Chapter 1, boundary lubrication depends on the presence of an oxide film on the metals' surfaces. Problems can occur in compressors handling inert or reducing gases, e.g., pure nitrogen and pure

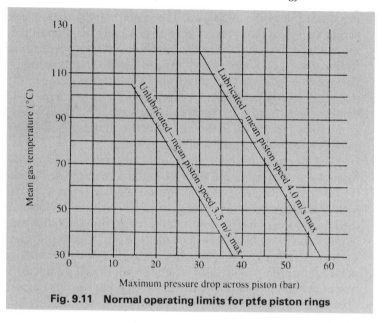

Fig. 9.11 Normal operating limits for ptfe piston rings

hydrogen, through loss of the oxide film. This can be overcome by introducing small amounts of moisture to the gas or via the lubricant.

Accurate fitting and running-in is particularly important with mechanical rod packings. A limit of 80°C is recommended for rod temperature. If this is reached the machine should be shut down and the rod allowed to cool; the process is repeated until the temperature no longer rises above the 80°C limit.

Carbon–graphite or filled ptfe piston rings and mechanical packings are used for unlubricated operation; these are somewhat specialized applications. Figure 9.11 gives guidance on normal operating limits for filled ptfe piston rings, both unlubricated and lubricated. More arduous conditions have to be discussed with the compressor and specialized piston and packing manufacturers.

In concluding this chapter on sealing it is worth reiterating that all rubbing seals are very much a problem in marginal lubrication, where factors like surface finish, minor amounts of impurity, wettability, temperature, and attention to detail become extremely important.

CHAPTER 10

Wear

Wear is complicated by the variety of wear mechanisms:

adhesive wear
abrasive wear
fatigue pitting

These frequently interact and can be dramatically affected by corrosion, i.e.,

corrosive wear

The basic wear relationship

wear = k. (load × sliding distance)

helps as an aid to understanding, but the specific wear coefficient, k, is so dependent on the conditions that designing against wear remains an empirical practice.

Wear is a phenomenon of outstanding technological significance; it is one of the three mechanisms by which equipment fails, the others being fracture and corrosion. The understanding of wear is, however, at a much lower stage of development than that of either friction or lubrication. Wear, the loss of material from a surface, except for the case of running-in wear, results in a deterioration. Loss of material can be the result of a variety of mechanisms such as abrasion, erosion and surface fatigue; in combination with corrosion the rate of loss can increase dramatically. In any practical situation the wear process can be a combination of the different wear mechanisms, the importance of the different mechanisms frequently changing during the wear process.

10.1   WEAR MECHANISMS

Abrasion is the result of hard particles cutting the surface and removing material: it can occur as a two-body process with two surfaces in sliding

177

contact, or a three-body process when hard particles get trapped between the sliding surfaces. Adhesion is the main process in friction arising from the force required to shear the small welds that occur at asperity contacts between two sliding surfaces. In time the repeated formation and breaking of such contacts leads to the detachment of loose particles. Fatigue is involved in adhesive wear, but it is convenient to separate surface fatigue caused by repeated mechanical stressing without the formation of micro-welds. Surface fatigue occurs in rolling contacts, for example in rolling bearings and at the pitch line in gears. It also occurs in hydrodynamically-lubricated bearings subject to alternating loads even if there is no contact between the journal and the bearing, the load being transmitted through the oil film. Surface fatigue is also caused by cavitation, which can be a wear mechanism at the tips of high-speed pump impellers and in bearings subjected to rapid load changes, which occur, for example, in engine bearings. Erosion caused by the impact of solid particles or liquid droplets is another cause of surface fatigue; small pieces break away through repeated stressing; an example is the wear at bends in pipelines carrying liquids containing solids.

In static conditions with a corrosive environment, a protective film can form on the surface, slowing up and even completely inhibiting the progress of further attack, but if the film is continuously removed by a mechanical wear process the rate of loss of material can be dramatic.

EFFECT OF CORROSION ON WEAR

Corrosion may form a protective film that reduces the rate of loss of material (e.g., action of load-carrying additives) – a form of *mild wear*. If the corrosion film is continuously removed by the wearing process the wear rate can increase by several orders of magnitude – a form of *severe wear* called *corrosive wear*.

It is relatively simple to distinguish between abrasive wear that gives rise to continuous score marks (Fig. 10.1), adhesive wear that causes scuffing or galling in which the scoring tends to be very rough and discontinuous (Fig. 10.2), and straightforward fatigue wear that results in the creation of pits (Fig. 10.3). However, when the various mechanisms listed above are

Fig. 10.1   Score marks on white metal journal bearing – abrasive wearing by hard contaminant particles. (Note the erosive wear in the oil groove resulting from failure to provide 'dirt grooves' – see Fig. 5.5)

examined more closely it is evident that they are not necessarily separate and independent: moreover, in any practical wear situation more than one wear process can be involved. Given the present stage of knowledge, it is probably better to approach practical wear problems from a phenomenological point of view rather than from the position of preconceived ideas about the different wear mechanisms.

In considering the wear of a rubbing surface it is helpful to make a distinction between those cases in which the two surfaces are similar to each other in hardness and those in which one surface is significantly harder than the other. The two conditions are similar to the adhesion and ploughing mechanisms put forward by Bowden and Tabor in their theory of sliding friction. A case in which there is a marked difference in hardness could be either an unlubricated bearing application such as a carbon–graphite bearing operating with a steel shaft, or the situation where an abrasive contaminant comes between the bearing surfaces. The situation under dry rubbing is comparatively straightforward, but wear also occurs under lubricated conditions; during starting and stopping, and also during running when there is some interruption of the oil film. This may be due to

Fig. 10.2 Scuffing in cylinder of reciprocating compressor caused by lubrication breakdown

asperities of rough surfaces penetrating the film or contaminant particles which may be present in the lubricant, being large enough to bridge the gap between the surfaces. Because of the intermittent nature of the surface contacts in the lubricated case, prediction of the wear behaviour is more complicated.

In general, as already indicated there are two distinct regimes of wear, usually referred to as 'mild' and 'severe' wear. In the former, the wear particles are very fine (0.01–0.02 μm) and there is comparatively little roughening of the worn surface. In 'severe' wear the wear particles are much larger and the surfaces become rough and torn in appearance, giving rise to what is usually termed scuffing or galling (Fig. 10.2). Although these terms are based on laboratory observations, they match the practical experience that the rubbing surfaces of machines wear either very

Fig. 10.3 Initial fatigue pit on ball bearing inner race

gradually over a long period of time, or very rapidly leading to seizure or catastrophic failure.

WEAR REGIMES

Mild wear   Wear particles fine – 0.01–0.02 μm
                    Surfaces polish
                    May be an acceptable operating condition.

Severe wear   Wear particles large – > 1 μm
                    Surfaces become rough and torn (*scuffing, galling*)
                    Unacceptable – leads to seizure or catastrophic failure

According to the adhesion theory of friction, welds are formed between high spots on the opposing surfaces and the friction arises from the force required to tear the welds apart. This suggests a mechanism for the creation of loose particles. Measurements of wear rates, however, show that a wear particle could not possibly form at every such encounter, and it appears more probable that we are again dealing with a fatigue mechanism, the particles finally breaking away after a number of such encounters. Where there is ploughing of a softer surface by a harder one, observations show that this can result in plastic deformation forming a wave round the indenting asperity or a cutting action when some of the softer surface is machined away. The former mechanism once again gives the conditions for fatigue, the latter immediately results in the formation of loose wear debris. This provides one explanation for the phenomena of mild and severe wear; the former occurring as the result of fatigue, the latter through cutting, though severe wear can also occur under adhesion conditions. Kraghelsky has shown that the separation between the processes of wear by fatigue and cutting is a function of the relative depth of penetration (the ratio of the depth of penetration to the radius of the indentor, taking this to be a sphere) and the coefficient of friction. The line in Fig. 10.4 shows the transition from deformation to cutting. Even when there is cutting, only a small proportion of the groove depth represents actual loss of material; the remainder is a consequence of plastic deformation. It can be seen that to obtain cutting not only must there be a hardness difference, but also the indenting asperities must have a small radius of curvature. If

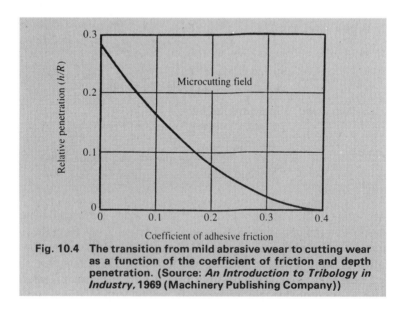

Fig. 10.4 **The transition from mild abrasive wear to cutting wear as a function of the coefficient of friction and depth penetration. (Source: *An Introduction to Tribology in Industry*, 1969 (Machinery Publishing Company))**

the asperity tips are broken off, or become smoother, as for example when the cylinder wall of an unlubricated reciprocating compressor with carbon piston rings becomes smoother during the initial running, the wear rate by cutting will decrease. Wear rate thus diminishes with running-in, which as already demonstrated in Chapter 1 results in the removal of the asperity peaks, both in dry rubbing and in the lubricated case. In the latter, this is because the chance of asperity penetration of the lubricant film is reduced and the probability of contaminant particles passing between the surfaces without bridging the gap is increased.

Mild wear is generally associated with the formation of a protective surface film generated by the rubbing process itself. In air these films are normally oxides and with metals there may be an appreciable electrical resistance at the contact. In severe wear the degree of metallic contact, and hence the opportunity for welding, is much greater. In general, mild wear occurs at low loads or speeds and severe wear at high loads or speeds, though a change may take place from severe to mild wear with increasing speed if the temperature generated by the increased friction loss results in the formation of a more stable protective oxide film on the surface. In the same way the chemical film formed by ep lubrication oil additives ensures mild wear on the breakdown of the oil film in a gear contact in place of the scuffing (severe wear) that would occur in its absence. Running-in, as already mentioned, is typical of mild wear, the surface becoming smoother as the tops of the protuberances are removed.

10.2 WEAR RELATIONSHIPS

It has been established experimentally that wear in many cases obeys the following simple relationship:

BASIC WEAR EQUATION

$V = kWS$

V = volume worn
k = wear coefficient
W = load
S = sliding distance

It will be seen that wear, like friction, is independent of the apparent area of contact. This relationship is obeyed for both cases of mild wear occurring under dry or lubricated conditions and severe wear under dry sliding, the value of the constant, k, being several orders of magnitude greater for severe wear. Figure 10.5 gives examples of combinations of metals where the wear rate under dry sliding conditions, expressed as the volume wear per unit sliding distance, is directly proportional to the load. Further, the

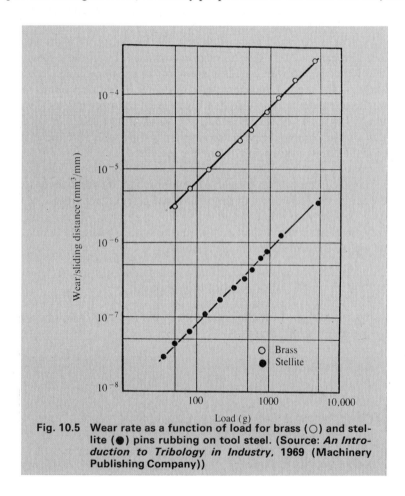

Fig. 10.5   Wear rate as a function of load for brass (○) and stellite (●) pins rubbing on tool steel. (Source: *An Introduction to Tribology in Industry*, 1969 (Machinery Publishing Company))

wear rate per unit load is inversely proportional to the hardness (Fig. 10.6). This suggests that the wear equation can be extended as follows

$$V = \frac{kWS}{p_m}$$

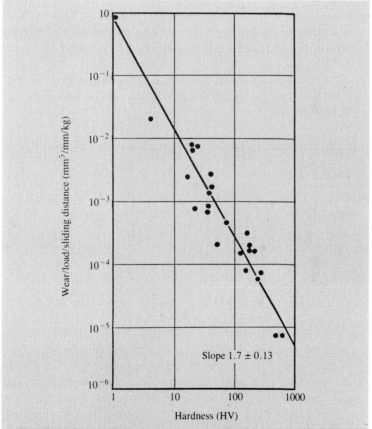

Slope 1.7 ± 0.13

Hardness (HV)

Fig. 10.6 Effect of hardness on wear rate of a number of materials against tool steel. (Source: *An Introduction to Tribology in Industry*, 1969 (Machinery Publishing Company))

The wear coefficient *k* varies by about six orders of magnitude, depending on the material combination, for a hardness range of only about three orders of magnitude. Thus *k* is a more important factor in determining the wear rate than hardness. From the practical point of view this is unfortunate, because while the hardness can be determined accurately, the exact physical nature of *k* is not clear and its value can only be determined by experiment. Moreover, its reproducibility is poor and the value obtained is extremely sensitive to the exact experimental conditions. This means that it is not possible to predict the wear rate of a combination of materials, and selection for wear resistance largely remains a matter of trial and error.

The wear equation can be modified by dividing through by the area of contact and replacing the sliding distance by the velocity multiplied by the time. We then obtain

LINEAR WEAR RATE

$$\frac{V}{A} = \frac{kWvt}{A}$$

and

$$\frac{l}{t} = kpv$$

where l/t is the linear (or depth) wear rate per unit of time and *p* the projected area loading. The wear rate is thus a function of the product *pv*, and for any given acceptable wear rate, a limiting *pv* value can be determined for a particular material (rubbing against a harder counterface). This is, however, an oversimplification as the equation is only valid for the range of *p* and *v* over which the wear mechanism does not alter, being limited at one end by the loading that causes plastic deformation and at the other by the speed that results in loss of mechanical properties through temperature rise caused by frictional heating. The *pv* factor can be taken as the first step in the selection of dry rubbing bearing materials, but as can be seen from Fig. 5.3 few materials obey the simple wear relationship over more than a narrow range of *p* and *v* values.

10.3 WEAR IN PRACTICE

It is worth examining in a little more detail the wear behaviour of two materials that are widely used in dry rubbing applications, for the further insight they throw on the wear process. These are carbon–graphite and reinforced ptfe, both of which roughly obey the *pv* wear relationship over a small range of *p* and *v* (Fig. 10.7). The wear rate of carbon–graphite is markedly affected by the roughness of the opposing surface. After a period of running, depending on its hardness, the countersurface tends to reach a value of about 0.1 μm R_a or more. (Although the graphite in carbon–graphite compacts is a soft, self-lubricating material, the carbon is

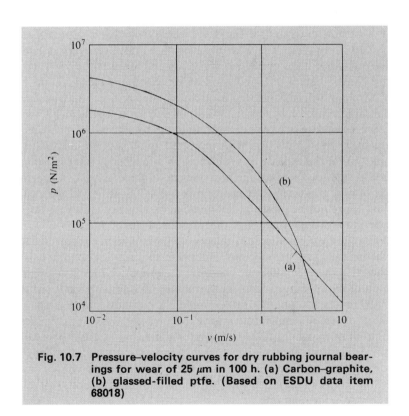

Fig. 10.7 Pressure–velocity curves for dry rubbing journal bearings for wear of 25 μm in 100 h. (a) Carbon–graphite, (b) glassed-filled ptfe. (Based on ESDU data item 68018)

hard enough to wear metal surfaces.) Under the run-in condition a typical carbon–graphite has a specific wear rate of approximately $0.5 \times 10^{-15} \, m^3/Ns$ (Fig. 10.7). This is an acceptable wear rate for a wide range of applications. However, it has been established that the low friction and good wear behaviour of carbon–graphite depends on the presence of adsorbed films of 'condensible gases' (e.g., water vapour and carbon dioxide) and in the absence of these the wear rate increases by a factor of about 100. Savage has shown that the vapour pressure of a particular gas to provide an effective lubrication film on graphite is a function of the molecular size (Fig. 10.8). For example, with water a vapour pressure 1/10th of saturation is necessary for lubrication, whereas with ethylene 1/350th of saturation is sufficient.

The specific wear rate of carbon and glass-filled ptfe is very similar to that of carbon–graphite (see Fig. 10.7, where the value is approximately $0.3 \times 10^{-15} \, m^3/Ns$). The wear mechanism is, however, very different. Filled ptfes have a low coefficient of friction and low wear only when rubbing against a film of ptfe, and good wear behaviour depends on the formation of a transferred film of ptfe on the countersurface. This transferred film is largely held on by mechanical forces and for it to adhere it is necessary for the surface to have a certain minimum roughness, probably in the range of 0.1 to 0.2 μm R_a. With increased roughness the film will still adhere, but at the expense of a higher wear rate when it is being formed; with too smooth a surface the film does not adhere as well. A practical range of roughnesses for use against ptfe is 0.4–0.6 μm R_a. In contrast to carbon–graphite, the presence of liquid films may give a high wear rate by interfering with the formation of the transferred film of ptfe.

While the wear equation can thus be useful in certain circumstances in predicting the life of components under defined operating conditions, it is implicit that there should be no change in the properties of the materials as a result of the rubbing process. Unfortunately this assumption is not true in many practical situations. For example, the hardness may decrease as a result of increased temperature arising from frictional heating or increase as a result of metallurgical transformations induced by the rubbing process. Furthermore, frictional heating may induce chemical reactions that modify the surface properties, for example by forming oxide films or modifying the nature of the oxide film. Any of these transformations will result in a change in the wear rate.

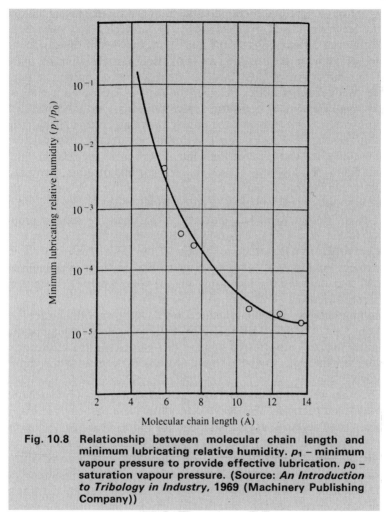

Fig. 10.8 Relationship between molecular chain length and minimum lubricating relative humidity. p_1 – minimum vapour pressure to provide effective lubrication. p_0 – saturation vapour pressure. (Source: *An Introduction to Tribology in Industry*, 1969 (Machinery Publishing Company))

So far we have been considering the wear that occurs between two sliding surfaces in solid contact. Wear also occurs with two surfaces in rolling contact, for example at the pitch line of gears and on the races or rolling elements of rolling bearings. Wear occurs here by the formation of pits, not by the gradual or rapid removal of the surface layers as occurs respectively in mild and severe adhesive wear between sliding surfaces.

Pitting is a fatigue process, the alternating stresses at the surface ultimately resulting in the breaking away of small particles from the surface and the formation of pits. Surface roughness is of particular importance. With rough surfaces most of the load is carried on a comparatively small number of opposing asperity contacts, with the result that very high stresses occur at these contacts; the asperities fail by fatigue and in this way the load is more uniformly distributed. Eventually pitting ceases when the stress at the contacts no longer exceeds the fatigue limit. This type of pitting occurs at the pitch line of gears. Provided it ceases it can be considered as running-in, but with rough surfaces there is a risk that the pitting may proceed to complete destruction of the gear teeth. Recent work has shown that the time to pitting is proportional to the ratio of the elasto-hydrodynamic oil film thickness at the contact to the initial surface roughness of the teeth. The most effective way to control pitting is to improve the surface finish, though increasing the oil film thickness by increasing the oil viscosity can help to retard the onset of pitting.

Rolling bearings are manufactured with a very smooth surface finish and this type of fatigue does not occur. Fatigue occurs in rolling bearings when the surfaces are stressed by hard dirt particles in the lubricant just larger than the elasto-hydrodynamic film thickness passing through the contacts.

The other wear situation that has to be considered is wear of a solid surface by loose particles. This may occur when these particles are trapped between solid surfaces, as just described for rolling bearings, and as happens in hydrodynamically-lubricated bearings when the lubricant is contaminated by solid particles small enough to enter the bearing with the lubricant feed, but large enough to bridge the gap at the point of minimum film thickness, or between the faces of a mechanical seal. It is also found when particles rub against a solid surface, as in a chute or at the working faces of earth-moving equipment, or impinge against a surface, as in the erosion of bearings or of pump or pipes handling fluids containing solids in suspension.

This field is probably best illustrated by some examples. With sliding bearings it is common practice to make one of the surfaces significantly softer than the other, e.g., white metal bearings operating against steel journals. The main intention is to confine any damage to one of the components, one that is a relatively simple component that can be readily replaced in the event of damage. An important advantage of using a soft bearing material is that it is able to embed hard contaminant particles and prevent them damaging the other surface. There is obviously a limit to the amount of contaminant that can be absorbed and, if this is exceeded, it is then possible for the softer component to wear the harder one by acting as a lap. This is not an uncommon situation with mechanical seals handling liquids containing solids, when the hard face frequently wears more than the softer one. To get wear of this nature, the contaminant particle must be harder than the hard face. A possible solution is to use two hard faces so that the contaminant particles are ground down and do not cause wear. For example, this is done in mechanical seals by using a combination of silicon carbide and tungsten carbide or two silicon carbide faces.

Another example where hardness appears to be the significant factor is the use of glass linings for chutes and hoppers handling coal and coke. These linings have a very satisfactory life, whereas mild steels wear very rapidly in this application. In complete contrast, a soft elastic material can give the same result: rubber linings are very effective in preventing wear of chutes handling catalyst pellets composed of minerals of extremely high hardness. Again, in shot-blasting booths rubber sheeting has an excellent life, whereas corrugated iron is very quickly penetrated. As mentioned above, rubber is also used successfully in cutless bearings lubricated by water containing silt, the rubber deforming elastically and allowing the silt particles to pass through without causing wear.

It is possible in the case of coal, referred to above, that corrosion may also be a relevant factor and that glass is largely effective because of its chemical inertness. For example, it has been found that a cast iron pipe handling a gypsum slurry effluent had a life at the bends of only three months, whereas a replacement pipe in PVC lasted for 2 ½ years. In the case of the cast iron pipe it is clear that the protective oxide film was continuously being removed and the pipe was wearing by a combined corrosion/erosion mechanism. PVC is much softer than the gypsum crystals and yet it did not wear to the same extent, though it was found that if the velocity in the pipe exceeded a certain value severe erosive wear occurred at the bends by a cutting mechanism. In a similar way the wear that occurs in slurry pumps has also been cured by the application of resin coatings to the wearing parts, particularly resin/slate powder mixtures.

A further example in which corrosion was clearly the dominant factor was in induced draft fans handling dirty, wet air containing sulphur dioxide. Here the life of steel impellers was only two to twelve weeks, whereas the application of a resin coating to the mild steel surfaces increased the life to over 1 ½ years. The resistance of resin coatings to wear by impingement from solid particles is shown by the fact that it is extremely difficult to clean the resin off resin-coated surfaces by shot-blasting.

Another analogous situation is when the component is forced through a more or less compact agglomeration of loose particles, as for example in earth moving, tilling, or, slightly differently, in coal cutting. Laboratory testing of the wear of metal specimens loaded against abrasive cloths has shown that for abrasives harder than the wear test specimen the wear resistance (i.e., the inverse of the wear rate) is approximately proportional to hardness (Fig. 10.9). The wear resistance is the same whether the material is annealed or fully work hardened, suggesting that work hardening occurs rapidly at the contact so that it is the work-hardened condition that is relevant. Where hardness is produced by metallurgical transformations, as for example by heat-treating steels, the wear resistance is still proportional to hardness, but at a lower rate.

Work on the wear of agricultural tools gives very similar results. Wear rate is dependent on the particular soil, and is influenced by the quantity and size of the hard abrasive present.

From the examples considered it is clearly an oversimplification to imagine that wear resistance is merely proportional to hardness. Corrosion can be a factor of overriding importance and the first requirement for reducing wear is that the materials selected should be chemically resistant to the environment. Once this has been achieved the two cases of wear at low and high rates of application of stress – the cases of rubbing and impingement wear – must be considered separately. Before considering these it is perhaps worth emphasizing the importance of corrosion in wear situations by some further practical examples.

A small injector was being used to pump a liquid into a high-pressure system. The plunger was lapped into the barrel of the pump. The injection rate was controlled by the speed of the pump and, as wear occurred, it was necessary to increase the speed to maintain the required rate and compen-

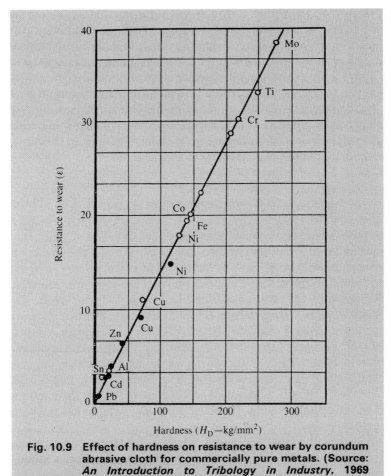

Fig. 10.9 Effect of hardness on resistance to wear by corundum abrasive cloth for commercially pure metals. (Source: *An Introduction to Tribology in Industry*, 1969 (Machinery Publishing Company))

sate for the leakage past the plunger as wear occurred, a replacement being required when the limiting speed of the equipment was reached. Some silica was present in the pumped liquid and it was suspected that this was responsible for the wear by abrasive action. Increasingly hard alloys were used, but to little effect. Examination of the plunger surface by electron microscopy suggested that corrosion was occurring. This was quite unexpected from the chemical point of view and was presumably a consequence of the high pressure. A change was made to a 13 percent chromium steel with a dramatic increase in the life of the pump, although this steel was appreciably softer than the materials previously used.

Another well known case where corrosion plays an important part in the wear process, is in the cylinders of diesel engines operating with sulphur-containing fuels. If the temperature in the cylinder is low enough for the sulphuric acid produced by the burning of the fuel to condense on the cylinder wall the wear rate increases significantly.

10.4   SOME GENERALIZATIONS

With two similar solid surfaces in rubbing contact, adhesive junctions are formed and torn apart, ultimately causing wear through the formation of loose particles that break away by fatigue. In this region wear can be of the 'mild' kind if the protective oxide film prevents excessive junction growth. If disruption of the protective oxide film occurs a change to the severe regime will occur. This transition is usually associated with an increase in temperature, giving a change in the nature of the oxide film and a decrease in the yield strength of the supporting substrate, though it can occur at low temperatures with aluminium, where the epitaxy (match of the crystal structure of the film to the underlying metal) is poor, so that hard oxide is rather brittle and fails to impede junction growth. (This is the reason why cold welding of aluminium is possible and why titanium has poor tribological properties.)

Where there is a large difference in hardness the asperities on the harder surface cut into the soft one, machining away part of the latter. As Kraghelsky has shown, the stress level at which this occurs is a function of the coefficient of friction at the contact. In the region of cutting, wear changes to the 'severe' type.

In a bearing in which it is not possible to prevent wear, it is preferable to restrict the wear to one of the two surfaces, which can then be made a simple component that is readily renewed when it is no longer serviceable. To get reasonable life, however, the wear should be in the 'mild' region. The chances of achieving this are improved by using a low friction material, hence the use of carbon–graphite or ptfe as bearings in situations where it is not possible to guarantee separation of the surfaces by a fluid lubricant film.

In situations where hard particles are present it is preferable to separate the surfaces by a fluid lubricating film. If contaminant particles are present, the risk of wear can be combated to some extent by reducing the contact stress, for example by using a material with a low modulus of elasticity, as shown for the case of rubber bearings lubricated with silty water. The alternative is to use either a soft material that will allow the hard particle to embed in it, the classic example being the use of white metal bearing alloys, or to use rubbing materials that are harder than the loose particles, for example, two hard faces in mechanical seals.

The second situation is where wear is caused by particle impingement or impact. Again, if the contact stress can be absorbed elastically there will be no wear. This principle is exploited in shot-blast booths, where rubber screens are used to absorb the high impact energy of the shot that would quickly erode metal screens. In contrast, in chutes where the impact energy is lower because of the reduced impact angle and lower particle velocity, glass linings can be effective. At higher stress levels where the elastic limit is exceeded, low modulus materials wear by cutting as they can no longer absorb the impact energy. Abrasive wear can be resisted by using sufficiently hard materials, e.g., hardened digger tips and tile linings in kilns, but fatigue failure can occur through repeated stressing, brittle materials failing more rapidly than those with high damping capacity.

An overriding factor that must be considered in all wear situations is the possibility of corrosion; if this is present the rate of wear will be considerably enhanced, for example in 'fretting corrosion', where the fretting wear particles react with moist air to form hard oxides.

It is clear that our understanding of wear is not sufficiently advanced for immediate solutions to be provided to wear problems. However, an appreciation of wear mechanisms and the knowledge of wear processes can provide a useful starting point. If nothing else, it is important to approach wear problems with an open mind and to discard some of our preconceived opinions. For example, would resin coating give longer life to tillage tools than the use of materials of increased hardness that are liable to fail by brittle fracture?

SELECTED BIBLIOGRAPHY

RECOMMENDED SOURCES OF INFORMATION

Although some of the following references would appear to be somewhat dated, they are in the author's opinion still the best available sources of useful information.

General

M. J. Neale (editor) *A Tribology Handbook: Bearings,* 1993 (Butterworth–Heinemann, Oxford).

M. J. Neale (editor) *A Tribology Handbook: Drives and Seals,* 1993 (Butterworth–Heinemann, Oxford).

M. J. Neale (editor) *A Tribology Handbook: Lubrication,* 1993 (Butterworth–Heinemann, Oxford).

M. J. Neale (editor) *A Tribology Handbook: Component Failures, Maintenance and Repair,* 1994 (Butterworth–Heinemann, Oxford).

Bearings – general

D. F. Wilcock and **E. R. Booser,** *Bearing Design and Application,* 1957 (McGraw-Hill, New York).

Dry rubbing bearings

ESDU Data Item 87007. *Design and material selection for dry rubbing bearings.* (ESDU – Engineering Sciences Data Unit, London).

Journal bearings

ESDU Data Item 66023. *Calculation methods for steadily-loaded pressure-fed hydrodynamic journal bearings.* (Nominally replaced by Data Item 84031, but in the author's opinion a better treatment that also covers journal bearings with central circumferential grooves.)

ESDU Data Item 84031. *Calculation methods for steadily-loaded axial groove hydrodynamic journal bearings.*

ESDU Data Item 77013. *Low viscosity process fluid lubricated journal bearings.*

ESDU Data Item 77034. *Journal bearing performance in superlaminar operation.*

Rolling bearings
Eschmann, Hasbargen, Wiegand and **Braendlein,** *Ball and Roller Bearings.*

ESDU Data Item 78032. *Grease life estimation in rolling bearings.*

ESDU Data Item 81005. *Designing with rolling bearings. Part 1: Design considerations in rolling bearing selection with particular reference to single-row radial and cylindrical roller bearings.*

ESDU Data Item 81037. *Designing with rolling bearings. Part 2: Selection of single row angular contact ball, taper roller and spherical roller bearings.*

J. A. Harris, *The Lubrication of Rolling Bearings,* 1966 (Shell Oil UK).

T. S. Nisbet and **G. W. Mullett,** *Rolling Bearings in Service,* 1978 (Hutchinson Benham).

Thrust bearings
ESDU Data Item 77030. *Calculation methods for steadily-loaded fixed inclined pad thrust bearings.*

ESDU Data Item 75023. *Calculation methods for steadily-loaded tilting-pad thrust bearings.*

Elasto-hydrodynamic lubrication
ESDU Data Item 85027. *Film thicknesses in lubricated Hertzian contacts (EHL). Part 1: Two dimensional contacts (line contacts).*

Gears
Nomenclature of Gear Tooth Wear and Failure, 1964 (American Gear Manufacturers' Association).

The Lubrication of Industrial Gears, 1964 (Shell International Petroleum Company).

Lubrication systems
ESDU Data Item 68039. *Guide to the design of tanks for forced circulation oil lubrication systems.*

ESDU Data Item 83030. *Selection of filter rating for lubrication systems.*

Mechanical seals
J. D. Summers-Smith (editor) *Mechanical Seal Practice for Improved Performance,* Second edition, 1993 (Mechanical Engineering Publications, London).

Vibration

C. **Jackson** *The Practical Vibration Primer,* 1979 (Gulf Publishing Company, Houston, London, Paris, Zurich, Tokyo).

VDI-Richtlinie. VDI 2056. *Criteria for assessing mechanical vibrations of machines.* (English translation Peter Peregrinus Ltd. 1971.)

Wear

M. E. **Peterson** and W. O. **Winer** (editors) *Wear Control Handbook,* 1980 (American Society of Mechanical Engineers).

INDEX

199